Cognitive Neuroscience of Decision-Making

There are many different types of decisions – from the important and life-changing to the mundane and everyday – but all are important for our functioning as humans. This book offers an accessible guide to the complex process of human decision-making, tailored for both undergraduate and postgraduate students. It combines recent research with real-life examples to provide a comprehensive understanding of the underlying biology of decision-making, its relationship to cognitive abilities such as working memory and attention, and its intersection with development. The book also explores applications and theories of decision-making, giving readers a broader perspective on the field. Presented in an accessible format with in-depth explanations, the work provides everything needed to build a strong basis of understanding of the underlying biology to the more complex topics of how decision-making develops and impacts on other behaviours. Discussion points are included throughout to encourage deeper reflection on the content covered.

Stacey A. Bedwell is a lecturer in psychology and neuroscience who has developed courses on the neuropsychology of decision-making at King's College London and the University of Cambridge. She was awarded her PhD in Neuroanatomy in 2015.

Cognitive Neuroscience of Decision-Making

Stacey A. Bedwell

King's College London

CAMBRIDGE
UNIVERSITY PRESS

![Cambridge University Press logo](CAMBRIDGE UNIVERSITY PRESS)

Shaftesbury Road, Cambridge CB2 8EA, United Kingdom

One Liberty Plaza, 20th Floor, New York, NY 10006, USA

477 Williamstown Road, Port Melbourne, VIC 3207, Australia

314–321, 3rd Floor, Plot 3, Splendor Forum, Jasola District Centre, New Delhi – 110025, India

103 Penang Road, #05–06/07, Visioncrest Commercial, Singapore 238467

Cambridge University Press is part of Cambridge University Press & Assessment, a department of the University of Cambridge.

We share the University's mission to contribute to society through the pursuit of education, learning and research at the highest international levels of excellence.

www.cambridge.org
Information on this title: www.cambridge.org/9781009407977

DOI: 10.1017/9781009407946

First published 2026

Cover image: Richard Newstead, The Image Bank via Getty Images

A catalogue record for this publication is available from the British Library

A Cataloging-in-Publication data record for this book is available from the Library of Congress

ISBN 978-1-009-40797-7 Hardback
ISBN 978-1-009-40796-0 Paperback

Contents

List of Figures   *page* vi
Glossary   vii

1 Introduction   1

2 Executive Function and Cognitive Control   3

3 The Prefrontal Cortex   22

4 Prefrontal Cortex Structure and Organisation   38

5 Neurotransmitters and Neurophysiology   54

6 Memory   66

7 Decision Networks   78

8 Decision-Making Styles and Models of Decision-Making   87

9 Development   97

10 The Role of Childhood Experiences in Decision-Making   112

11 Decision-Making Deficits   127

12 Implications of Decision-Making   144

Index   152

Figures

2.1 Baddeley's (1986) model of working memory   *page* 6
2.2 Anderson's (2002) model of executive function   9
2.3 Example of Stroop experiment stimuli   13
3.1 Location of the prefrontal cortex   23
3.2 Phineas Gage   24
4.1 The original Brodmann's map, as published in Brodmann (1909)   39
4.2 Sensory homunculus   42
4.3 Processing hierarchy   47
4.4 Processing hierarchy   49
5.1 Synaptic transmission   59
6.1 Atkinson and Shiffrin's (1968) multi-store model of memory   72
7.1 The location of PFC connections in temporal cortex   79
7.2 Comparison of input and output connections from the PFC   80
7.3 Reciprocity of PFC connections   82
11.1 Example Iowa gambling task   129
11.2 Example Eriksen flanker task   130

Publisher's note: The e-book edition of this title, like the print editions, contains colour. For those e-reader devices and applications that cannot display colour, the colour material is available in pdf format as an online resource: https://cambridge.org/bedwell

Glossary

Abstract thinking: The ability to conceptualise thoughts, ideas, concepts and stimuli that are not linked to concrete objects or instances.

Anterior: The position towards the front of an organism, the opposite to posterior.

Attentional control: The ability to regulate and manage one's attention on a given stimulus or task, filtering out distractions.

Cell migration: The movement of cells from one physical location to another within an organism.

Cognitive control: The ability to regulate thoughts, actions and processes, and integrate and manipulate information for high-order processes.

Cognitive flexibility: The ability to adapt and adjust one's thinking to a given scenario.

Cortex: The outermost layer of the brain, often called the cerebral cortex or neocortex. Divided into two hemispheres, and further into four lobes.

Cytoarchitecture: The organisation and arrangement of cells.

Dorsal: The upper position on an organism, opposite to ventral.

Executive function: A set of high-order complex processes including planning, goal direction and decision-making, that integrate information from multiple systems.

Free will: To have conscious control over one's own actions and decisions.

Hippocampus: A structure in the brain, located in the temporal lobe. Highly associated with memory.

Homologous: Where two entities are similar to one another, for example similar in structure or traits.

Inhibitory control: Synonymous with response inhibition. The ability to inhibit automatic responses.

Lamination: The layered organisational structure of the cortex.

Lesion: An area of damage, injury or disease.

Lobotomy: A surgical procedure that involves severing connections in the prefrontal cortex.

Medial: The position towards the midline of an organism. Opposite to lateral.

Phylogenetic development: The evolutionary developmental history of a species or group of organisms.

Posterior: The position towards the back of an organism. Opposite to anterior.

Prefrontal cortex: The most anterior pole of the mammalian brain.

Psychopathology: The scientific understanding of psychological disorders.

Reciprocal: Mutual or complementary relationship. Reciprocal connections transmit information bidirectionally.

Thalamus: A structure located between the cortex and midbrain. Involved in sensory and motor signalling, memory, and conscious and cognitive functions.

Top-down processing: Cognitive processing by which existing knowledge influences how stimuli are perceived.

Topographically mapped connections: Brain connections that are spatially organised.

Ventral: The lower-side position on an organism. Opposite to dorsal.

1
. . . .

Introduction

As humans, we make countless decisions every day. Some decisions are important and life-changing, but most are mundane decisions we make multiple times – but decisions, nonetheless. For instance, this morning you likely decided to get out of bed. Getting out of bed wasn't the only decision that took place in that instance; you decided to open your eyes, to sit up, to put one leg in front of the other and perhaps to walk to the bathroom. Each one of these actions required a decision, even if it was not something you consciously thought about. Later, you may have decided between toast and cereal for breakfast and considered what you should wear to work that day after looking at the weather forecast. These are more consciously considered decisions; you might have actively thought about your previous experiences of not wearing a coat and it raining to inform your decision to wear a coat that day. Through your day you may well make some decisions that seem more important or influential than these, perhaps deciding what answer to give in a quiz at university that will impact your grade or deciding to apply to a specific job. There are many types of decisions, but all are important for our functioning as humans.

Over the past century we have gained a great deal of understanding of human decision-making, and the decision-making of other species. We now have greater knowledge than ever in terms of how decision-making contributes to a range of high-order processes, how decision-making can be influenced through both biological and environmental factors, and to an extent what goes wrong in the context of decision-making to contribute to the development of a range of psychological and neurological disorders.

There is, however, a long way to go. It remains to be established exactly what experiences and individual differences in decision-making development led to the vast range of decision-making behaviours seen among individuals and between cultures. Although we can be confident that some disorders involved decision-making dysfunction, we do not yet have the detailed understanding of the how and why, that is vital for the future developments of treatments, interventions and prevention.

Through the chapters of this book, you will develop an understanding of decision-making from a neuropsychological perspective, beginning with the gross anatomy of the brain regions involved in the decision process and working right through to questioning whether we really do have free will in the decisions we make.

2
• • • • • • •

Executive Function and Cognitive Control

2.1 What Is Executive Function?

Defining executive function is not a straightforward task, because there is a lack of consensus among experts as to how it should be explained. This is somewhat a reflection of how complex executive function is, and how much there remains to understand. It is naturally difficult to define a concept that cannot be seen or easily measured, as is the case with executive function.

Executive function is an umbrella term used to describe a set of high-order cognitive abilities related to complex tasks such as meeting goals, social interaction, problem-solving and planning ahead. Executive functions include cognitive flexibility, working memory, reasoning and inhibitory control (Cristofori et al., 2019). Importantly, executive function is not an isolated cognitive process but an integration of different processes and functions, associated with multiple brain regions and systems. The idea of integration is key when it comes to understanding executive function. In fact, it is not really a function but a method of combining and utilising a range of different functions. The complex cognitive functions we collectively refer to as executive function are heavily associated with, but not limited to, the prefrontal cortex. It is understood that the prefrontal cortex integrates information from the environment and other brain regions and systems to facilitate executive functions (Barbas, 2009).

Executive function encompasses many high-order skills that require inhibitory control, working memory and cognitive flexibility. These important elements begin developing in early childhood and continue through a prolonged period of development until they are considered fully developed around the age of twenty-five (Diamond, 2013). Through executive functioning, these high-order cognitive skills interact to facilitate complex behaviours related to goal attainment and goal-directed thought (Berthelsen et al., 2017). The delay in development of these high-order cognitive functions can be seen in the exhibited behavioural development of children and adolescents in related functions; for example, greater risky decision-making in teenagers is a result of the ongoing development of the complex network connectivity associated with complex decision-making (see Chapter 3).

Having established that executive function is not actually a function but a mechanism of integration and interaction of different high-order functions, Diamond (2013) suggested that the complex abilities enabled by executive function consist of three main elements: inhibition, working memory and cognitive flexibility.

2.2 Inhibitory Control

Inhibitory control is one of the vital elements that make up executive function. The ability to inhibit certain actions, behaviours or impulse responses enables us to focus on a given task.

The important ability of inhibitory control can be described as having control over automatic or natural urges by utilising other abilities such as attention and emotional control to refocus and respond to stimuli in a thought-out, appropriate way. Inhibition is required for us to be able to supress unwanted responses and process internal thoughts successfully (Lee & Chao, 2012). This process can involve considering the consequences of actions in a particular situation, where the urge to respond in a certain way might not have the most desirable consequences. A good example of this is responding to an itchy mosquito bite. The natural urge may be to scratch it, but considering past experiences and what the consequences of scratching might be, using inhibitory control, we might decide not to scratch.

This ability plays an important role when it comes to being able to work towards a goal, or plan ahead. Having inhibitory control is vital for self-regulation of emotions and control of behaviour in social situations (Lopez et al., 2021). Inhibitory control is vital in our ability to hold back from many behaviours ranging from saying inappropriate things in social situations, not interrupting people when they are speaking (behaving in socially acceptable way during a conversation) to not scratching insect bites when we feel the urge to. We are not always consciously aware when we are inhibiting behaviours (Ingerslev, 2020).

The ability to use inhibitory control begins to develop quite early in life. In fact, evidence shows one-year-old babies to exhibit some level of inhibition (Kochanska et al., 2010). The ability continues to develop into early adulthood, and declines again into older age, along with other executive functions (Ferguson et al., 2021).

Research evidence shows that people with a greater ability in inhibitory control often show greater academic attainment (Whedon et al., 2020), show better regulation of emotions (Fox & Calkins, 2003) and often make healthier lifestyle choices (Adams et al., 2017). Lower than average inhibitory control has been linked to ADHD and addiction (Kamarajan et al., 2005; Kenemans, 2015; Luijten et al., 2011; Suarez et al., 2021; Yongliang et al., 2000). This may be a result of lower inhibitory control impacting ability to regulate behaviours.

2.3 Types of Inhibitory Control

Inhibitory control can apply to cognition, behaviour, emotion and motor levels:

Cognitive Inhibition

- Being able to focus attention on a given task, and not other distracting stimuli. For example, paying attention to a lecture and not the football game outside the lecture hall window.

Behavioural Inhibition

- The ability to control urges to behave or respond in a certain way, where our previous experience and social understanding tell us it would be inappropriate or have negative consequences.

Emotional Inhibition

- The ability to regulate emotions.

Motor Inhibition

- Our ability to control motor behaviours; for example, remaining in the lecture hall seat, even though you are uncomfortable and want to stand up and move around.

2.4 Working Memory

Working memory refers to a limited capacity element of the memory system, which allows the temporary storage and manipulation of information for cognitive processing. Baddeley's (1986) model of working memory (Figure 2.1) describes four key components: the phonological loop, visuospatial sketchpad, executive control system and episodic buffer (more on models of memory in Chapter 6).

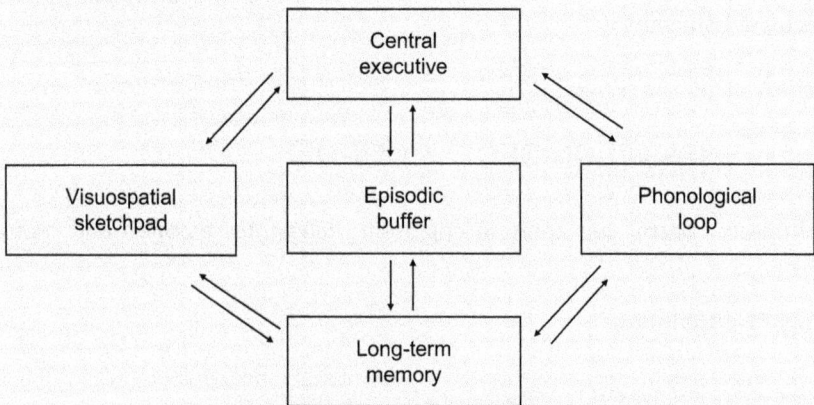

Figure 2.1 Baddeley's (1986) model of working memory.

- The phonological loop deals with language-based information; for example, the process of reading and listening to others speak.
- The visuospatial sketchpad handles visual information, as well as spatial information. For instance, it is involved in your ability to solve a jigsaw puzzle.
- The central executive does not deal with information itself, directly, but rather coordinates the processing of information by the visuospatial sketchpad and phonological loop. It is described as an attentional system that somewhat supervises the subsidiary slave systems. For instance, hearing directions (utilising the phonological loop) and visualising the route you need to take (utilising the visuospatial sketchpad) – the central executive is thought to allow the integration of information from these two elements of the system.
- The episodic buffer is a third slave system that is thought to provide temporary integration of information before it is retrieved, reflected and manipulated by the central executive.

Working memory is a vital component of numerous cognitive functions. It is especially important in executive function. The process of integrating information from all over the brain, from multiple networks, based on varied experiences and memories, relies on working memory. Working memory as a function allows us to integrate all of this information in the present. Working memory could be considered as a mental workspace, where information is actively manipulated.

2.5 Cognitive Flexibility

The third key component of executive function, according to Diamond (2013), is cognitive flexibility. This refers to our ability to adapt our thinking and behaviour to different situations and in response to different stimuli and events. Cognitive flexibility requires the ability to shift attention and adjust responses in real time. This can often require mentally considering alternative outcomes, based on past knowledge and experience and, where problem-solving is involved, considering different strategies and how they might play out. This all requires the ability to think abstractly.

The prefrontal cortex is important for cognitive flexibility because it allows the coordination of different information and varied cognitive processes, and allows for task switching (Friedman & Robbins, 2022). Disorders associated with changes to the prefrontal cortex such as schizophrenia are also linked to deficits in cognitive flexibility (Wang et al., 2022).

2.6 Alternative Conceptualisation of Executive Function

As it is such a complex and difficult-to-define concept, there are naturally many variations in the way executive function has been explained or conceptualised. Much in the same way multiple models of memory were developed in the last century, many variations have been developed in an attempt to explain and model executive function. In comparison to Diamond, Hughes and Ensor (2011) conceptualised executive function as a broad umbrella term for a wide range of complex cognitive abilities, all of which coordinate or supervise goal-oriented actions in some way. According to Anderson (2002), the concept of executive function can be broken down into four key components (Figure 2.2).

2.6.1 Attentional Control

Often better known as concentration, attentional control refers to an individual's ability to control which stimuli (internal or external) they pay attention to at a given moment; an ability to concentrate on a specific task at hand. This is important for focusing cognitive resources on an important task. Attention is often the first stage of a more complex cognitive task. Attentional control is commonly viewed as being either goal-driven or stimulus-driven. Vecera et al. (2014) argued that it is more accurately described as driven by experience. Specifically, driven by experience of the current stimuli. If the individual has little experience of the stimulus, the attention will be stimulus-driven. If the individual has some experience of the stimulus, the attention is more goal-driven. The authors state that attention becomes more goal-driven as experience grows.

Attentional control can be further categorised into four distinct types of attention.

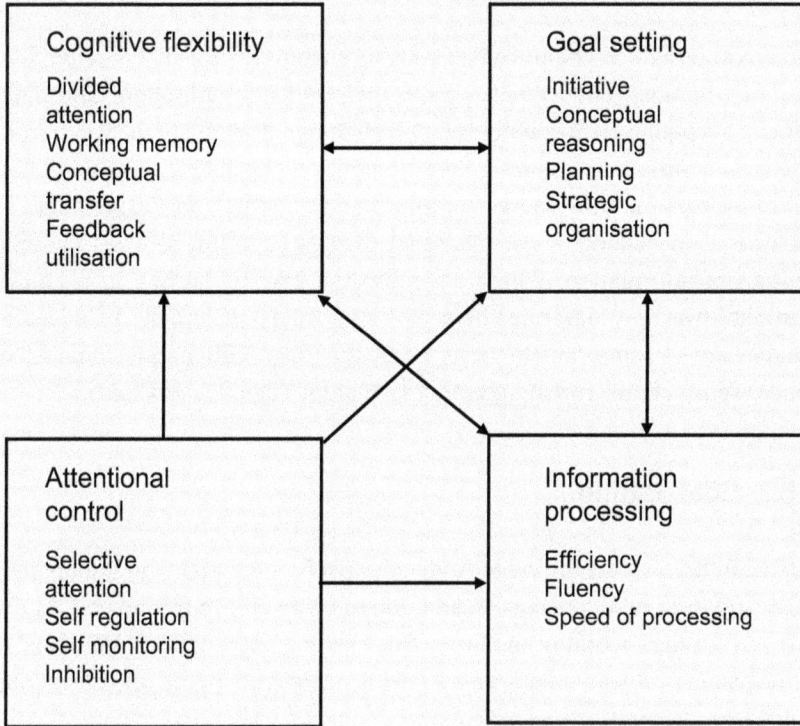

Figure 2.2 Anderson's (2002) model of executive function.

- *Selective attention* refers to an ability to select one or more elements of a stimulus to pay attention to, whilst essentially filtering out other, perhaps irrelevant, elements.
- *Sustained attention* is an ability to maintain focus on one given task, or on one given stimulus for an extended period. During sustained attention other distractor stimuli are not given any attention.
- *Alternating attention* is a similar concept to task switching. It is an ability to move focus between two or more different stimuli or tasks requiring different cognitive resources.
- *Divided attention*, in comparison to alternating attention, is the ability to maintain focus on two or more tasks simultaneously, with equal attention. Both tasks require different cognitive resources.

2.6.2 Cognitive Flexibility

Anderson described cognitive flexibility as encompassing divided attention, working memory, conceptual transfer and feedback utilisation. The main benefit of cognitively flexibility is task switching – the ability to switch from one task or stimulus to another. Being cognitively flexible allows an individual to process multiple sources of information and use past experiences in their decision-making. It also allows for divided attention between multiple sources of information. Where an individual may be lacking in cognitive flexibility, we may describe them as being rigid and unable to adapt to changes. It is important to note, as is the case with many domains of executive function, working memory is vital for cognitive flexibility.

2.6.3 Goal Setting

The third domain according to Anderson involves the ability to initiate new ideas and form new concepts, in addition to being able to plan ahead. Again, this all requires working memory. Goal setting incorporates elements of strategising, problem-solving and strategic behaviours based on considering outcomes and previous experiences.

2.6.4 Information Processing

The final domain, information processing, essentially brings the others together. Executive functions are complex and demanding in terms of neurochemistry and anatomy. For executive functioning to work, the underlying processing of information must be highly efficient, and fast. The neural networks that allow such complex cognition to run smoothly have to be extremely high performing for the outputs expected. Deficits in this domain can result in delayed responses or hesitancy is decisions where processing is not fast enough.

Executive function has a prolonged development, beginning to develop through childhood and continuing into early adulthood. The initial development of executive function, and the individual differences seen in elements of executive function, are thought to be dependent on both genetic and environmental factors. According to Fishbein et al. (2019), there is a link between home environment in childhood and specific abilities in executive function. There is a correlational relationship between positive home environment and executive function performance. The same are also linked to reduced behavioural problems.

According to Anderson (2002) some domains of executive function, namely attentional control, can be observed as early as infancy, whereas other domains (cognitive flexibility, goal setting and information processing) are more delayed in their development, showing a rapid period of development in late childhood (around eight years). This aligns with knowledge surrounding executive function performance, whereby younger children can show basic abilities, such as making straightforward decisions, but cannot perform more demanding executive tasks such as more complex decision-making. Anderson (2002) stated executive control can be seen at the beginning of adolescence.

2.7 Abstract Thinking

In the context of executive function, abstract thinking refers to an ability to form and conceptualise ideas that are not concrete, or that cannot be perceived in the environment, such as planning for an event in the future or designing a building that does not yet exist. Abstract thinking is important in our advancement as a species, because it allowed us to go beyond the limitations of our perceivable environment. Essentially, abstract thinking allows us to mentally represent things that do not physically exist or have not happened. It allows us to engage in creative thought processes and allows for the components of executive function: inhibitory control and cognitive flexibility.

2.8 Cool and Hot Executive Functions

Zelazo and Muller (2002) proposed a two-dimensional model of executive function, suggesting that cognitive processes that fall under the umbrella term of executive function can be further categorised into either cool or hot executive functions. Hot executive functions involve emotion regulation and motivation, whereas cool executive functions are considered more abstract, not related to emotion. There is some debate in the literature regarding what does and does not classify as a cool/hot executive function. For instance, some suggest abilities including theory of mind, emotional regulation and empathy are considered hot executive functions (De Luca & Leventer, 2008; Happaney et al., 2004), whereas others suggest these abilities, although complex, are not associated with hot executive functions (Zelazo et al., 2005). This is just one example of how difficult it is to define and describe in detail such a complex concept as executive function.

2.9 Top-Down Processing

A traditional model of hierarchical organisation states that information travels from the higher-order regions to the most primary regions – from association cortices to secondary cortices to primary cortices; for example, prefrontal cortex ➔ secondary motor cortex ➔ primary motor cortex.

In basic cognition, processing is described as either top-down or bottom-up. Bottom-up processing is considered the simpler of the two. This involves interpreting the world based on information gathered by the senses, at the time. On the other hand, top-down processing involves using pre-existing knowledge and experiences to interpret the world. Top-down processing is thought to be important in decision-making because it allows us to use pre-existing knowledge to quickly make decisions.

2.10 The Stroop Effect

A clear example of top-down processing can be seen in the famous Stroop phenomenon. In this commonly used experimental task, subjects are shown a series of words presented in coloured ink (Figure 2.3). The words either match the coloured ink; for example, 'blue' printed in blue ink, or do not; for example, 'red' printed in blue ink. The object is to name the colour of the ink, not read the typed word. People tend to be much slower in reading the words and make more mistakes when the colours do not match – in the case of 'red' printed in blue ink.

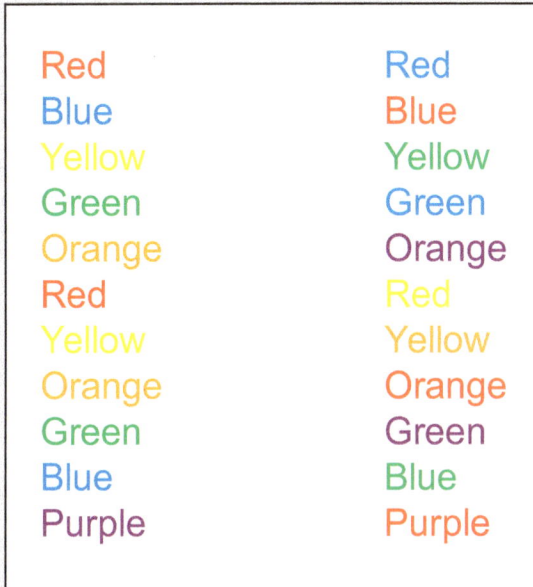

Figure 2.3 Example of Stroop experiment stimuli.

The difference in abilities seen between the two conditions of the Stroop task – the Stroop effect – can be explained by the concept of top-down processing. Due to prior experience, most people will recognise the typed word before thinking about the colour ink it is printed in.

2.11 Evidence against Top-Down Processing of Executive Functions

Based on our knowledge of other brain regions, including complex regions like the visual cortex, it would have been sensible to assume the prefrontal cortex and its associated executive functions work in very much the same established patterns. And on a broad scale, this is the case for the prefrontal cortex. However, recent evidence exploring the fine-scale anatomical connectivity of the prefrontal cortex in rats suggests prefrontal cortical connectivity is organised differently to the traditional hierarchical model.

Recent anatomical evidence (Bedwell et al., 2017) has revealed an unusual circuit when viewed in terms of a traditional hierarchical model (Botvinick,

2008; Fuster, 2001). Studies in rats consistently identify direct connections between primary somatosensory cortex and prefrontal cortex, which would not be expected from a traditional view of information flow, that is, sensory cortex → higher sensory cortex → association cortex → prefrontal cortex → secondary motor cortex → motor cortex. The meaning of this connection in terms of behaviour is not yet clear; however, the knowledge of its existence provides a valuable insight into the overall organisation of the prefrontal cortex and how this may feed into complex cognitive processes like decision-making.

DISCUSSION POINT

Suggest an explanation for the differences observed in prefrontal cortex circuitry by Bedwell et al. (2017).

2.12 What Is Cognitive Control?

The terms executive function and cognitive control are sometimes used synonymously, and often confused. Although they are interchangeable in some circumstances, it is important to differentiate the two. Executive function refers to the higher-level processes that enable cognitive control. Cognitive control can be defined as an ability to have control over responses to given situations and stimuli. This can be applied in a range of contexts such as problem-solving, decision-making and goal direction.

2.13 Historical Perspective on Cognitive Control

Thoughts about cognitive control began to appear among scholars around the end of the nineteenth century and the beginning of the twentieth. Early ideas focused on the concept of attention, such as selective attention and inhibition (Dewey, 1896; James, 1890). Although contemporary understanding of cognitive control and what it encompasses goes far beyond attention, this understanding of the important of attentional systems was a vital first step in the developments in understanding that came over the next 100 years and into the twenty-first century.

The cognitive revolution of the mid twentieth century led to many advances across cognitive science. Prominent psychologists at the time, such as George Miller, led the way in the development of numerous cognitive theories to explain behaviours. Miller (1956) offered explanations for the concept of cognitive control, integrating processes of attention, memory and problem-solving. In the 1950s Broadbent made a distinction between automatic and controlled processes, leading to the idea of selective attention. It was the development of Baddeley's (1986) model of working memory that paved the way for a very structured approach to trying to understand the complexities of cognition, through the introduction of the concept of the central executive, a cognitive component responsible for the integration and coordination of multiple processes.

Towards the end of the twentieth century and the beginning of this century, researchers Posner, Fuster and Shallice began to lay the groundwork for what we now understand to be cognitive control. They carried out a lot of work exploring executive functions. For instance, Posner and Petersen (1990) were the first to suggest a specific executive branch of the established attentional system at the time. They suggested it is this executive branch that allows us to focus attention on a given task at hand. Shallice et al. (2008) built on the idea of the executive branch with a description of a supervisory system, which allows us to focus attention on the task at hand by essentially overriding automatic responses.

2.14 Technological Advances

Our contemporary understanding of cognitive control has been greatly influenced by technological advances in the field of cognitive science over the past century. The introduction of neuroimaging methods such as functional magnetic resonance imaging have allowed investigations and visualisations of cognitive function that could not have been imagined in the past. It was such advances in imaging methods that led to understanding the importance of the prefrontal cortex in executive function and as a key region for cognitive control (E. K. Miller & Cohen, 2001).

2.15 Seven Features of Cognitive Control

Based on knowledge built through the pioneering research of the late twentieth and early twenty-first centuries, our current understanding suggests cognitive control can be described by seven features:

1. Necessary to deal with novel tasks.
2. Extends beyond current environment.
3. Necessary to initiate new sequences of behaviour.
4. Prevents inappropriate responses.
5. Rapid task switching.
6. Monitors performance, corrects errors.
7. Sustained attention.

2.16 Importance of Working Memory in Cognitive Control

It is understood that past knowledge or recently obtained information can influence current behaviour; this is relevant to cognitive control as well as simpler behaviours. Working memory is a key function for a wide range of cognitive processes but is especially important when it comes to cognitive control. Working memory maintains information in the present; it keeps information active to make decisions or inhibit behaviours. A key aspect of cognitive control is abstract representation; our ability to represent information that is not currently available in reality – being able to mentally picture something you are not currently experiencing. Working memory is important for abstract representation to be successful; it allows for the integration of information from past experiences (abstract representation) and information from the present, to inform whatever decision is to be made or cognitive control to be implemented.

2.17 Explaining Cognitive Control

As a complex, hard-to-define concept, cognitive control and executive control have attracted several variations in models attempting to explain it in a digestible form.

2.17.1 Dynamic Filtering Theory

Shimamura (2000) described four elements that comprise executive control.

1. Selecting
2. Maintaining
3. Updating
4. Rerouting
   - Selecting is the ability to select specific stimuli or information, and focus attention, like the suggestion made by Posner.
   - Maintaining is the maintenance of information in short-term memory – highlighting the importance of working memory.
   - Updating is how Shimamura described the ability to reorganise and refocus information whilst it is being maintained in short-term memory. This is important for manipulating information and integrating to inform decisions.
   - Rerouting is the ability to task switch, to change from one cognitive process or a response to another. This links to earlier ideas of resisting natural urges.

Shimamura (2000) also stated that cognitive control involves a dedicated filtering region, the prefrontal cortex. This region has a great control over varied cognitive processes from across the brain, and it is here that the complex abilities described by the four-element model (updating, rerouting) are enabled and regulated.

2.17.2 Integrative Theory

Miller and Cohen's (2001) integrative theory takes many of the same familiar concepts, with a focus on the prefrontal cortex. They proposed that cognitive control stems directly from patterns of activity in the prefrontal cortex, of which consistent maintenance represents goals and the drive to achieve them. It is the prefrontal cortex that guides inputs of information, integrates information from past and present experiences from different regions across the brain and mediates actions in response to environmental stimuli. Importantly, the prefrontal cortex provides guidance to other cortical regions via top-down processing – which brings to question how evidence against consistent top-down processing in the prefrontal cortex (Bedwell & Tinsley, 2018) fits in to what we understand about cognitive control.

2.18 What Is the Difference between Executive Function and Cognitive Control?

They are often used interchangeably, but do not necessarily mean the same thing. One main difference is the control element; not all executive functions involve control. So, although most cognitive control involves executive functions, not all executive functions are involved in cognitive control. A good example of this is time perception – this is considered an executive function but does not involve any element of cognitive control.

Key Points

- Executive function describes a set of high-order cognitive abilities.
- Three main elements of executive function are inhibition, working memory and cognitive flexibility.
- Working memory is important in executive function; integration of information from all over the brain relies on working memory.

REFERENCES

Adams, R. C., Lawrence, N. S., Verbruggen, F., & Chambers, C. D. (2017). Training response inhibition to reduce food consumption: Mechanisms, stimulus specificity and appropriate training protocols. *Appetite, 109*, 11–23. https://doi.org/10.1016/j.appet.2016.11.014

Anderson, P. (2002). Assessment and development of executive function (EF) during childhood. *Child Neuropsychology, 8*(2), 71–82. https://doi.org/10.1076/chin.8.2.71.8724

Baddeley, A. (1986). *Working memory.* Clarendon Press/Oxford University Press.

Barbas, H. (2009). Prefrontal cortex: Structure and anatomy. In L. R. Squire (Ed.), *Encyclopedia of neuroscience* (Vol. 7, pp. 909–918). Academic Press.

Bedwell, S. A., Billett, E. E., Crofts, J. J., & Tinsley, C. J. (2017). Differences in anatomical connections across distinct areas in the rodent prefrontal cortex. *European Journal of Neuroscience, 45*(6), 859–873. https://doi.org/10.1111/ejn.13521

Bedwell, S. A., & Tinsley, C. J. (2018). Mapping of fine scale rat prefrontal cortex connections: Evidence for detailed ordering of inputs and outputs connecting the temporal cortex and sensory-motor regions. *European Journal of Neuroscience, 48*(3), 1944–1963. https://doi.org/10.1111/ejn.14068

Berthelsen, D., Hayes, N., White, S. L. J., & Williams, K. E. (2017). Executive function in adolescence: Associations with child and family risk factors and self-regulation in early childhood. *Frontiers in Psychology*, *8*, 903. https://doi.org/10.3389/fpsyg.2017.00903

Botvinick, M. M. (2008). Hierarchical models of behavior and prefrontal function. *Trends in Cognitive Sciences*, *12*(5), 201–208. https://doi.org/10.1016/j.tics.2008.02.009

Cristofori, I., Cohen-Zimerman, S., & Grafman, J. (2019). Executive functions. In M. D'Esposito & J. H. Grafman (Eds.), *Handbook of clinical neurology* (Vol. 163, pp. 197–219). Elsevier. https://doi.org/10.1016/B978-0-12-804281-6.00011-2

De Luca, C. R., & Leventer, R. J. (2008). Developmental trajectories of executive functions across the lifespan. In V. Anderson, R. Jacobs, & P. J. Anderson (Eds.), *Executive functions and the frontal lobes: A lifespan perspective* (pp. 23–56). Taylor & Francis.

Dewey, J. (1896). The reflex arc concept in psychology. *The Psychological Review*, *3*(4), 357–370. https://doi.org/10.1037/h0070405

Diamond, A. (2013). Executive functions. *Annual Review of Psychology*, *64*, 135–168. https://doi.org/10.1146/annurev-psych-113011-143750

Ferguson, H. J., Brunsdon, V. E. A., & Bradford, E. E. F. (2021). The developmental trajectories of executive function from adolescence to old age. *Scientific Reports*, *11*(1), 1382. https://doi.org/10.1038/s41598-020-80866-1

Fishbein, D. I., Michael, L., Guthrie, C., Carr, C., & Raymer, J. (2019). Associations between environmental conditions and executive cognitive functioning and behaviour during late childhood: A pilot study. *Frontiers in Psychology*, *10*. https://doi.org/10.3389/fpsyg.2019.01263

Fox, N. A., & Calkins, S. D. (2003). The development of self-control of emotion: Intrinsic and extrinsic influences. *Motivation and Emotion*, *27*, 7–26. https://doi.org/10.1023/A:1023622324898

Friedman, N. P., & Robbins, T. W. (2022). The role of prefrontal cortex in cognitive control and executive function. *Neuropsychopharmacology*, *47*(1), 72–89. https://doi.org/10.1038/s41386-021-01132-0

Fuster, J. M. (2001). The prefrontal cortex – an update: Time is of the essence. *Neuron*, *30*(2), 319–333. https://doi.org/10.1016/s0896-6273(01)00285-9

Happaney, K., Zelazo, P. D., & Stuss, D. T. (2004). Development of orbitofrontal function: Current themes and future directions. *Brain and Cognition*, *55*(1), 1–10. https://doi.org/10.1016/j.bandc.2004.01.001

Hughes, C., & Ensor, R. (2011). Individual differences in growth in executive function across the transition to school predict externalizing and internalizing behaviors and self-perceived academic success at 6 years of age. *Journal of Experimental Child Psychology*, *108*(3), 663–676. https://doi.org/10.1016/j.jecp.2010.06.005

Ingerslev, L. R. (2020). Inhibited intentionality: On possible self-understanding in cases of weak agency. *Frontiers in Psychology*, *11*, 558709. https://doi.org/10.3389/fpsyg.2020.558709

James, W. (1890). *The principles of psychology* (Vol. 1). Henry Holt.

Kamarajan, C., Porjesz, B., Jones, K. A., Choi, K., Chorlian, D. B., Padmanabhapillai, A., Rangaswamy, M., Stimus, A. T., & Begleiter, H. (2005). Alcoholism is a disinhibitory disorder: Neurophysiological evidence from a Go/No-Go task. *Biological Psychology, 69*(3), 353–373. https://doi.org/10.1016/j.biopsycho.2004.08.004

Kenemans, J. L. (2015). Specific proactive and generic reactive inhibition. *Neuroscience & Biobehavioral Reviews, 56*, 115–126. https://doi.org/10.1016/j.neubiorev.2015.06.011

Kochanska, G., Woodard, J., Kim, S., Koenig, J. L., Yoon, J. E., & Barry, R. A. (2010). Positive socialization mechanisms in secure and insecure parent–child dyads: Two longitudinal studies. *Journal of Child Psychology and Psychiatry, 51*(9), 998–1009. https://doi.org/10.1111/j.1469-7610.2010.02238.x

Lee, Y., & Chao, H. (2012). The role of active inhibitory control in psychological well-being and mindfulness. *Personality and Individual Differences, 53*(5), 618–621. https://doi.org/10.1016/j.paid.2012.05.001

Lopez, R., Cosme, D., Werner, K. M., Saunders, B., & Hofmann, W. (2021). Associations between use of self-regulatory strategies and daily eating patterns: An experience sampling study in college-aged females. *Motivation and Emotion, 45*, 747–758. https://doi.org/10.1007/s11031-021-09903-4

Luijten, M., Littel, M., & Franken, I. H. (2011). Deficits in inhibitory control in smokers during a Go/NoGo task: An investigation using event-related brain potentials. *PLoS One, 6*(4), e18898. https://doi.org/10.1371/journal.pone.0018898

Miller, E. K., & Cohen, J. D. (2001). An integrative theory of prefrontal cortex function. *Annual Review of Neuroscience, 24*, 167–202. https://doi.org/10.1146/annurev.neuro.24.1.167

Miller, G. A. (1956). The magical number seven, plus or minus two: Some limits on our capacity for processing information. *Psychological Review, 63*(2), 81–97. https://doi.org/10.1037/h0043158

Posner, M. I., & Petersen, S. E. (1990). The attention system of the human brain. *Annual Review of Neuroscience, 13*, 25–42.

Shallice, T., Stuss, D. T., Picton, T. W., Alexander, M. P., & Gillingham, S. (2008). Mapping task switching in frontal cortex through neuropsychological group studies. *Frontiers in Neuroscience, 2*, 79–85.

Shimamura, A. P. (2000). The role of the prefrontal cortex in dynamic filtering. *Psychobiology, 28*, 207–218. https://doi.org/10.3758/BF03331979

Suarez, I., De los Reyes Aragón, C., Grandjean, A., Barceló, E., Mebarak, M., Lewis, S., Pineda-Alhucema, W., & Casini, L. (2021). Two sides of the same coin: ADHD affects reactive but not proactive inhibition in children. *Cognitive Neuropsychology, 38*(5), 349–363. https://doi.org/10.1080/02643294.2022.2031944

Vecera, S. P., Cosman, J. D., & Vatterott, Z. J. (2014). The control of visual attention: Toward a unified account. In J. Wixted & S. W. Lindsay (Eds.), *The psychology of learning and motivation* (Vol. 60, pp. 303–347). Academic Press.

Wang, Y., Hu, X., & Li, Y. (2022). Investigating cognitive flexibility deficit in schizophrenia using task-based whole-brain functional connectivity. *Frontiers in Psychiatry, 13*. https://doi.org/10.3389/fpsyt.2022.1069036

Whedon, M., Perry, N., & Bell, M. A. (2020). Relations between frontal EEG maturation and inhibitory control in preschool in the prediction of children's early academic skills. *Brain and Cognition, 146*(2), 105636. https://doi.org/10.1016/j.bandc.2020.105636

Yongliang, G., Robaey, P., Karayanidis, F., Bourassa, M., Pelletier, G., & Geoffroy, G. (2000). ERPs and behavioral inhibition in a Go/No-Go task in children with attention-deficit hyperactivity disorder. *Brain and Cognition, 43*(1–3), 215–220.

Zelazo, P. D., & Muller, U. (2002). Executive function in typical and atypical development. In U. Goswami (Ed.), *Blackwell handbook of childhood cognitive development* (pp. 445–469). Blackwell.

Zelazo, P. D., Qu, L., & Müller, U. (2005). Hot and cool aspects of executive function: Relations in early development. In W. Schneider, R. Schumann-Hengsteler, & B. Sodian (Eds.), *Young children's cognitive development: Interrelationships among executive functioning, working memory, verbal ability, and theory of mind* (pp. 71–93). Erlbaum.

3

· · · · · · ·

The Prefrontal Cortex

3.1 Importance of the Prefrontal Cortex

The final frontier in the understanding of human cognition and behaviour, vital in the execution of complex processes, the human prefrontal cortex (PFC) is a common topic of discussion in contemporary brain research. Most neuroscientists and psychologists today would acknowledge the importance of this somewhat illusive brain region, but this was not always the case. Until relatively recently, the PFC was assumed, at least in Western science, to hold no real function, and this is understandable, given our knowledge now surrounding the abstract and difficult-to-observe functionality associated with prefrontal regions. The functionality of the PFC is not something that could historically be measured or observed, in the same way the function of motor or sensory cortices could be seen; stimulating a specific region of the motor cortex for instance initiates movement in the hand, making it obvious what the role of that area of cortex is. Similar investigations in the PFC would result in no observable effect in this way, giving the impression it does not control anything. Of course, with modern science and using indirect methods of measurement, we know this is far from the truth.

3.2 Defining the Prefrontal Cortex

The frontal lobe is separated from the parietal lobe by the central sulcus, and from the temporal lobe by the lateral sulcus. The frontal lobe is the

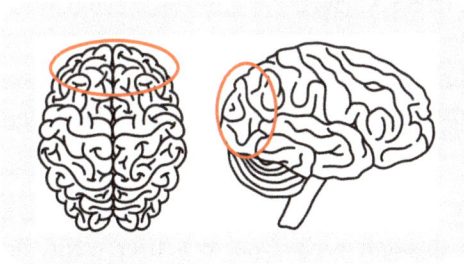

Figure 3.1 Location of the prefrontal cortex.

Source image @marina_ua / iStock / Getty Images Plus.

area of cortex generally associated with high order and executive functioning. This includes decision-making, forward planning, goal direction and social inhibition. The frontal lobe is also the area most linked to personality. The most anterior region of the frontal lobe of the mammalian brain is known as the PFC. The area of cortex often referred to as the PFC (Figure 3.1) has been defined by a granular layer IV. Rose and Woolsey (1948) defined the PFC as the projection target of the mediodorsal nucleus of the thalamus. Others have described the PFC as a region that exhibits stronger connections with a specific part of the thalamus, the mediodorsal nucleus. This is in comparison to weaker connections with other nuclei (Uylings et al., 2003). This definition, based on anatomical connectivity, is commonly utilised across mammalian species, including humans. Although somewhat limiting to anatomy and excluding function, this consistent and measurable definition is important in the continued study of the PFC.

3.3 Historical Ideas

Ideas surrounding the role of the area of the brain we know today as the PFC began long before modern science. The ancient Greeks were one of the first civilisations thought to identify the brain as a centre of consciousness. Philosophers including Plato and Aristotle famously discussed the role of the mind and of reasoning in human behaviours. Although the ancient Greeks may not have explicitly labelled the anatomical region we today know as the PFC, they had begun to relate complex behaviours and decisions to the brain.

3.4 The Legacy of Phineas Gage

Some of the earliest understanding of the PFC and its role in human functioning came from the famous case study of Phineas Gage, depicted in Figure 3.2 (Harlow, 1868).

Phineas Gage was a miner. In 1848, on a normal day at work, he was using a tamping iron to prepare to blast through rock. Due to a user error, he accidentally caused an explosion, sending the iron bar straight through his skull and through his frontal lobe, in through the left side of his face, near his eye and out through the top of his head, through the top frontal area of his skull. The rod passed straight through his frontal lobe. Perhaps surprisingly, Phineas survived the accident with little obvious impairment in terms of brain function. Recollections report that he was standing, speaking and walking just minutes after the accident. In the medical report from the same

Figure 3.2 Phineas Gage

PD-US-expired.

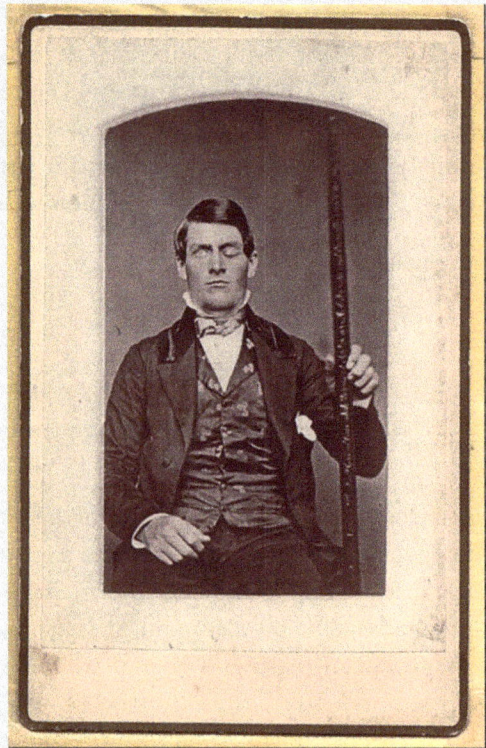

day, the doctor described brain tissue being visible from an opening in the top of his skull.

He was blinded in his left eye but was able to function in his daily life in a way that was considered fairly normal. For instance, his motor control, memory and sensory processing were all as expected. However, after the accident, friends and relatives supposedly started to notice differences in his personality. They described that before the accident Phineas had been very hard-working and responsible but afterwards had become quite disrespectful and had difficulty carrying out plans. We now understand these impairments to be associated with PFC damage. In fact, it was this unique opportunity to observe behaviour following such extensive prefrontal damage that enabled experts at the time to start to identify cognitive functions involving the previously mysterious region. This case paved the way for future research further implicating personality with the frontal lobe.

Phineas Gage lived a further eleven years, dying as a result of severe epilepsy at the age of thirty-six.

The case of Phineas Gage is somewhat contested today, as there is limited evidence to support the claims made about his change in behaviour. We should also acknowledge, with contemporary understanding of the effects of trauma, that in the process of receiving anatomical damage to the PFC, Phineas Gage went through a very traumatic experience. This experience may have had a lasting impact, regardless of the extensive physical damage. Despite the controversy, the case of Phineas Gage was a very important one, that opened scientific research into what was previously assumed a largely non-functional area of the human brain.

3.5 Lobotomy

Until as recently as the mid twentieth century, treatment of mental illness was based primarily on locking patients away in mental institutions or asylums. Although now often considered one of medicine's great failures, direct damage to the frontal lobes as treatment for symptoms of schizophrenia and depression, known as frontal lobotomy, became very popular between 1930 and 1950.

The procedure, introduced in the early twentieth century, is usually considered gruesome by today's standards. The practice of lobotomy sought to alleviate the emotional elements of mental illness, which had been associated with the frontal cortex. It started out by drilling holes into the skull (Moniz, 1935, as cited in Gross & Shafer, 2011) and was further developed by US neurologists Freeman and Watts (1936, as cited in Caruso & Sheehan, 2017) into the more familiar procedure of an ice-pick-like instrument called a leucotome hammered through the eye sockets (Faria, 2013). Sadly, many people died as a result of lobotomy procedures, but lobotomy was still long considered effective because it had the desired effects of reducing hard-to-manage symptoms.

Although we may consider the use of lobotomy a tragedy, there are important lessons about prefrontal function to be learnt from the effects seen in lobotomy patients. Many showed a reduction in anxiety and fewer emotional outbursts, as was the desired effect. Patients also often showed an impaired ability in complex tasks, for example delayed spatial responses. Other notable effects included a loss of goal-oriented behaviour, cognitive control and long-term purposeful behaviour. The loss of these complex functions leads to a greater understanding of the role of frontal regions.

3.6 Deep Brain Stimulation

In modern society, the practice of lobotomy has disappeared, but its legacy lives on. Deep brain stimulation utilises the historical knowledge gained from lobotomy. In this procedure, implanted electrodes are used to directly stimulate specific target regions of the brain, used today to treat many cases of major depressive disorder.

3.7 Lesion Studies

Despite the negative effects of lobotomy, many studies in the first half of the twentieth century went on to conclude the PFC was cognitively silent. That being, removal or damage seemingly had no lasting effect on behaviour (Hebb, 1949). Many experts continued to believe that the PFC was not needed for normal human functioning. Although advances in neuroimaging over the past century have led to numerous breakthroughs in the understanding of this illusive region, many important findings came from lesion studies in both humans and animals. Extensive research in anatomy, physiology and function have now confirmed what we understand to be a very cognitively important brain region.

Early lesion studies in primates, conducted in the 1930s (Jacobsen, 1935) demonstrated the role of the dorsolateral PFC (DLPFC) in working memory – a function vital to several high-order cognitive functions, including decision-making. Lesions to the DLPFC in monkeys lead to inabilities in specific tasks related to working memory. Similar lesion studies have identified the role of the PFC in anxiety, empathy, ambition and social control. There have been great advances in our understanding of PFC function over the past fifty years; however, the fine detail of the underlying structure remains largely unknown.

Despite its established functional importance, there remains relatively little knowledge in contemporary neuroscience regarding how the prefrontal region is organised and structured. The current lack of structural knowledge and understanding poses difficulties in understanding PFC functions, like decision-making, and to a great extent prevents progress in the understanding, prevention and treatment of prefrontal-associated deficits and psychological disorders that are thought to involve decision-making dysfunction (more on this in later chapters).

3.8 The Prefrontal Cortex across Species

It is important to recognise that the PFC, although very developed and clearly important in humans, is not a human-specific region. Many of the high-order functions we associate with the region are much more developed in humans, but we are not the only species that can make a decision or plan ahead.

The PFC has been recognised in many species, so it is by no means a uniquely human structure or region. But importantly, there is a level of controversy among neuroscientists in terms of whether the region we refer to as the PFC in different species is directly comparable to the region we describe in humans and other primates; for example, whether we can directly compare rats and humans. Brodmann, a very famous neuroanatomist, known for his cytoarchitectural maps of the brain, created a representation of the PFC in monkeys with a clear resemblance to that in humans. This established at least a reasonable comparison between us and close primate relatives. There is a consistent opinion among experts to this day that the architecture, or structure, of the PFC in humans is very similar to that seen in other primates. Research also shows similarities between primates and rats. This is good news when it comes to anatomical research, which often employs animal models, specifically rats.

The similarities between the human PFC and other mammalian species is not confined to structure but extends to function as well. The similarities between species are very important when it comes to understanding the PFC in humans, because it allows us to project findings from non-human animals onto humans. This is useful when exploring fine details which cannot yet be studied in the human brain but can in animal models.

The species most genetically like humans is the bonobo. Although neuroscience research does not tend to be carried out on bonobos (there are strict laws on experiments involving apes in the UK and Europe), many studies do involve the use of other primates such as macaque monkeys.

3.9 Controversy in the Generalisability of Animal Studies

It is accepted that something representing a PFC is recognisable in a range of species. However, without a single definition of what a PFC is, there does remain some controversy when it comes to comparison of the PFC in animals and humans, especially when considering its presence in non-primates such as rats and mice. Brodmann's (1909) representation of the PFC in monkeys greatly resembles that of the human (although some homologies are unclear). There is a consistent opinion among scholars that the architecture of the human PFC is largely similar to that of other primates (Petrides & Pandya, 1994; Semendeferi et al., 1998), so comparisons between humans and other primates are largely accepted.

Based on anatomical connections to other regions identified in primates, orbital prefrontal regions have been reliably identified in rats, cats and rabbits (Rose & Woolsey, 1948). Similarly, common features have been confirmed between specific regions of the PFC in primates and rats, based on anatomical features (Goldman-Rakic, 1988; Uylings & van Eden, 1990). So there is evidence for anatomical commonalties, despite obvious differences, for example in terms of size of the PFC and in terms of how advanced executive functions are in different species. There is also thought to be functional similarities between PFC regions across species; for example, comparable anatomical regions in both primates and rats are known to represent the same functions (Uylings et al., 2003).

Importantly for the use of animal models in research, the rat PFC is understood to share many commonalities with the human PFC. Although the human PFC is clearly more advanced in terms of function, the basic input and output connections from the region are very much the same (Holmes & Wellman, 2009). The common features between species mean that studying the complexity of prefrontal anatomy and function in rat models can be mostly applicable to humans, allowing for greater advances in our understanding than any currently possible human studies would allow.

Although it is clear and accepted by experts that rats possess much of the PFC properties found in humans and other primates, there remains no dispute that the rat PFC is not directly comparable to that of a human. Rats are known to lack the granular PFC that we see in primates, which is thought to be a structural property resultant of evolution – so only found in more highly evolved species. Despite this obvious anatomical difference, as well as the observable differences in advanced cognitive abilities, neuroscientists have claimed that based on similarities in the effects of lesions to the PFC, as well as the known similarities in connectivity, that at least the medial aspect of the frontal cortex in rats is homologous to the granular lateral PFC in primates (Kolb, 2007; Seamans et al., 2008). It is valuable to note that this region in humans is associated with psychopathology, psychosis, sociopathy, depression and addiction. This knowledge makes the rat a useful model on which to build an understanding of underlying PFC organisation, structure, connectivity and physiology, possibly giving rise to clearer understanding of various disorders in humans.

Although rats are commonly utilised to research anatomy, there is no dispute that rats lack the same prefrontal development as humans; it's clear that humans are more advanced as a species and we can visually see we have a much larger PFC, both in size and the proportion of our brain. It has been noted that rats lack the granular PFC found in primates, which is thought to be a product of evolution. Despite this, neuroanatomists have claimed, based on similarities in the effects of lesions, anatomical and physiological similarities that the medial aspect of the PFC in rats is homologous to the missing granular lateral PFC that we can see in humans and other primates. So, the medial PFC of the rat is said to be comparable to human PFC regions associated with psychopathology, psychosis, depression and addiction. This makes the rat a useful model on which to build an understanding of PFC organisation, which is comparable to areas of interest in humans. This explains why you see so much PFC research in rats and how it can be generalised to humans, even when we say the human PFC is so much more evolved.

3.10 Evolution and Phylogenetic Development

Considered from an evolutionary perspective, the mammalian PFC is thought to be the newest brain region. That is, for humans and other mammals alike, it is the part of the brain that evolved most recently. Evidence shows that through phylogenetic development – for example, if we compare the less evolved modern-day cat to the more evolutionarily advanced modern-day human – the PFC increases in size (Brodmann, 1909). This is in terms of the literal physical volume and mass, as well as in size relative to the animal and its brain. So, the percentage of a cat's brain taken up by the PFC is less than the percentage of a human brain taken up by the PFC. The more evolutionarily advanced an animal is, the greater amount of their total brain mass is taken up by the PFC. The PFC is recognisable across all mammalian species but reaches its largest size in primates. In humans, this amounts to approximately 30 per cent of total brain mass (Brodmann, 1909, 1912). Looking back through human evolution, the PFC increased in size over time, more so than any other cortical region (Preuss, 2000). Evolution tells us that natural selection acts on specific brain structures and regions that are advantageous in some way. The somewhat disproportionate size of the PFC in the human brain, taking up one-fifth of the brain, suggests that its function must be very

advantageous. This may be in terms of survival, adaptation to the environment or competition with other species.

As well as occupying a substantial area of the brain, the human PFC is the
final cortical region to be considered to reach maturity. The PFC and its
associated functions are usually considered fully developed around the age
of twenty-five (Casey et al., 2008; Sowell et al., 1999), which is relatively late
when compared to other complex regions like the motor and visual cortex.
This prolonged development means the complex cognitive functions associated with the region, including decision-making, goal direction, inhibition
and forward planning, are not fully developed until our mid-twenties. The
delayed development of the PFC in comparison to other regions could be a
result of highly complex connectivity, in the region itself and between the
PFC and other cortical and subcortical structures. Such complex connectivity likely takes longer to develop than the arguably simpler connectivity
elsewhere in the brain.

DISCUSSION POINT

Is research using animal models of value for investigating decision-making?

3.11 Prefrontal Cortex Function

The localisation of function refers to linking a specific physical part of the
brain to a specific function. This is an ongoing interest in neuroscience and
psychology research, not only in the PFC but across the brain. Localisation
of function is relevant not only for the complex high-order functions like
decision-making, that are associated with the PFC, but for a lot of processes
or functions that are linked to regions across the whole brain. When considering localisation, it is important to remember that a specific function can
rarely be pinpointed to one specific anatomical location in the brain. For
instance, there is not a specific location for making a decision; the process of
making a decision employs a number of different complex networks that
recruit neurons from all over the brain, to work together in carrying out a
complex task. For many cognitive processes and tasks, however, there are

some key regions and structures that are highly active. The localisation of function is important in enabling us to understand individual functions in more detail. The finer detail in which we can begin to understand about certain functions inevitably leads to the need to understand the underlying structure and functional mapping in more detail. This is the case for understanding of decision-making and the role of the PFC.

Once assumed to hold no real or specific function, neuroscientists now understand that the PFC is actually very important for a range of processes. The main reason it was assumed to have no real function for such a long time is because the functions associated with it are abstract; they are not things you can physically see like in the sensory or motor cortex, where for example you can see a limb move when a specific area of cortex is active.

We now understand that the PFC is one of the most highly connected regions within the human cortex (Fuster, 1999). Connections to the PFC can be seen from other brain regions that are known to be involved in functions like attention, cognition, action, emotion, reward, expectation, processing of outcomes, arousal, sleep, memory, movement, inhibition and distraction (Gehring & Willoughby, 2002; Goldman-Rakic, 1996; Schnider et al., 2005; Stuss et al., 1986; Stuss & Levine, 2002). It is clear that the PFC is involved in a wide range of important functions.

Considered historically, the frontal pole of the brain has long been connected with anxiety and emotional outbursts. This link was initially established by the removal of the PFC, a procedure known as frontal lobotomy. These early observations were an important starting point for what we now understand about the complex PFC.

The location of the PFC at the most anterior pole of the brain, at the front of the forehead just behind the skull, along with its large volume, make it quite vulnerable to potential damage and injury. Damage through traumatic brain injury can often result in impairments to the complex processes and functions associated with the PFC. Case studies of PFC injury have been beneficial in revealing the importance of the region in personality and emotion, as well as involvement in anxiety, ambition and self-control. Think back to the case of Phineas Gage.

The consensus from current understanding is that the PFC acts as an integrative association area. That is, it does not hold its own specific function but rather integrates information from all over the brain. The PFC is widely recognised as having the ability to represent information which is not currently available in reality. We call this abstract processing or representation. Researchers propose that abstract representation of information and events in the PFC enables the region to guide us in thought, action and emotion. Current observations in neuroscience research have shown the PFC to be highly associated with executive functions such as forward planning, decision-making and goal-directed behaviour.

Following the revelations from case studies like Phineas Gage, early studies into the human brain in contemporary neuroscience quickly began to identify functions associated with the region we now describe as the PFC. Investigations linking anatomy and function demonstrated that the PFC is involved in anxiety and emotional outbursts. Removal of the PFC clearly showed a reduction in such symptoms but also impairments in complex tasks (Jacobsen, 1935). Taken with the knowledge gained from observations, case studies and the outcomes of lobotomies, these early experimental findings formed an important basis for our current understanding of how the PFC is involved in emotion, reward and some of the most complex human functions. We have come a long way since assuming the region held no real function just a few hundred years ago.

Injury to the front of the head can often lead to impairments in the functions we now understand to be reliant on the PFC and its connectivity with other cortical and subcortical regions. Case studies continue to be valuable in piecing together our knowledge of this highly complex region and have revealed the importance of the region in personality and emotion (confirming the observations made in the Phineas Gage case), as well as involvement in anxiety, empathy, ambition and social control (Fuster, 2008; Goldman-Rakic, 1995).

Today, in contrast to early views of unimportance, the PFC is widely recognised by neuroscientists, psychologists and practitioners as being vital in our human ability to represent information which is not available in the present reality – abstract representation (Bowman & Zeithamova, 2018; Fuster et al., 2000; Goldman-Rakic, 1996; Levy, 2024). From an evolutionary perspective, it is thought that it is the ability for abstract representation, enabled by the complexities of the PFC, that allow us to be guided in complex thought processes that enabled us to advance so significantly as a species. Examples of how this is advantageous are planning to build a shelter or deciding based on learned knowledge from experience and from others informing us how to avoid predators or how to start a fire. Working memory forms a vital component of this ability, in the integration of information from across the brain, and allowing for information from past and present to be used together in the current decision-making process (Fuster, 1990; Fuster et al., 2000).

There have been great advances over the past century in our understanding of the PFC and its role in the human experience. However, despite many important steps in improving knowledge, there is a long way to go. The literature shows there remains a large amount of controversy in conclusions made from current observations, and this is largely because there is still a lot left to be explained in finer detail. This is especially true when compared to our level of understanding of other cortical regions, like the visual cortex, of which we have a much greater undisputable knowledge. At this point, experts have accepted that the PFC is critical to advanced human function and was pivotal in our evolution. Our complex PFC may still hold many mysteries for modern science, but we can conclude that it allows us as a species to implement complex cognitive functions. The underlying complexity of the PFC allows us to plan actions in the future, to change actions according to environmental cues, to achieve a goal. Through the PFC we can evaluate events as they happen, based on prior learning (Fuster et al., 2000). All of these highly complex processes require very complex circuitry, more complex than other brain regions.

Key Points

- The PFC is an integrative region, in that it integrates information from all over the brain.
- The PFC is recognised in many species – it is not unique to humans. However, it is most advanced in humans.
- The functions associated with the PFC are largely abstract – they are not functions that can be physically seen or easily measured.

REFERENCES

Bowman, C. R., & Zeithamova, D. (2018). Abstract memory representations in the ventromedial prefrontal cortex and hippocampus support concept generalization. *Journal of Neuroscience, 38*(10), 2605–2614. https://doi.org/10.1523/JNEUROSCI.2811-17.2018

Brodmann, K. (1909). *Beschreibung der einzelnen Hirnkarten IV. kapitel in Vergleichende Lokalisationslehre der Grosshirnrinde* [Description of individual brain maps. Chapter IV in *Localisation in the cerebral cortex*]. Verlag von Johann Ambrosias Barth.

Brodmann, K. (1912). Neue Ergebnisse über die vergleichende histologische Localisation der Grosshirnrinde mit besonderer Berücksichtigung des Stirnhirns [New results on the comparative histological localisation of the cerebral cortex with special consideration of the frontal lobe]. *Anatomischer Anzeiger* (Suppl.), *41*, 157–216.

Caruso, J. P., & Sheehan, J. P. (2017). Psychosurgery, ethics, and media: A history of Walter Freeman and the lobotomy. *Journal of Neurosurgery, 24*(3), E6. https://doi.org/10.3171/2017.6.FOCUS17257

Casey, B. J., Jones, R. M., & Hare, T. A. (2008). The adolescent brain. *Annals of the New York Academy of Sciences, 1124*, 111–126. https://doi.org/10.1196%2Fannals.1440.010

Faria, M. A. (2013). Violence, mental illness, and the brain – A brief history of psychosurgery: Part 1 – From trephination to lobotomy. *Surgical Neurology, 4*(49). https://doi.org/10.4103%2F2152-7806.110146

Fuster, J. M. (1990). Behavioral electrophysiology of the prefrontal cortex of the primate. *Progress in Brain Research, 85*, 313–323.

Fuster, J. M. (1999). Synopsis of function and dysfunction of the frontal lobe. *Acta Psychiatrica Scandinavica* (Suppl.), *395*, 51–57.

Fuster, J. M. (2008). *The prefrontal cortex* (4th ed.). Academic Press.

Fuster, J. M., Bodner, M., & Kroger, J. K. (2000). Cross-modal and cross-temporal association in neurons of frontal cortex. *Nature, 405*(6784), 347–351. https://doi.org/10.1038/35012613

Gehring, W. J., & Willoughby, A. R. (2002). The medial frontal cortex and the rapid processing of monetary gains and losses. *Science, 295*(5563), 2279–2282. https://doi.org/10.1126/science.1066893

Goldman-Rakic, P. S. (1988). Topography of cognition: Parallel distributed networks in primate association cortex. *Annual Review of Neuroscience, 11*, 137–156. https://doi.org/10.1146/annurev.ne.11.030188.001033

Goldman-Rakic, P. S. (1995). Architecture of the prefrontal cortex and the central executive. *Annals of the New York Academy of Sciences, 769*, 71–83. https://doi.org/10.1146/10.1111/j.1749-6632.1995.tb38132.x

Goldman-Rakic, P. S. (1996). The prefrontal landscape: Implications of functional architecture for understanding human mentation and the central executive. *Philosophical Transactions of the Royal Society of London. Series B, Biological Sciences, 351*(1346), 1445–1453. https://doi.org/10.1098/rstb.1996.0129

Gross, D., & Schafer, G. (2011). Egas Moniz (1874–1955) and the "invention" of modern psychosurgery: A historical and ethical reanalysis under special consideration of Portuguese original sources. *Journal of Neurosurgery, 30*(2), E8. https://doi.org/10.3171/2010.10.FOCUS10214

Harlow, J. M. (1868). Recovery from the passage of an iron bar through the head. *Publications of the Massachusetts Medical Society, 2,* 327–347.

Hebb, D. O. (1949). *The organization of behavior: A neuropsychological theory.* Wiley and Sons.

Holmes, A., & Wellman, C. L. (2009). Stress-induced prefrontal reorganization and executive dysfunction in rodents. *Neuroscience and Biobehavioral Reviews, 33,* 773–783. https://doi.org/10.1016/j.neubiorev.2008.11.005

Jacobsen, C. F. (1935). Functions of frontal association area in primates. *Archives of Neurology and Psychiatry, 33*(3), 558–569. https://doi.org/10.1001/archneurpsyc.1935.02250150108009

Kolb, B. (2007). Do all mammals have a prefrontal cortex? *Evolution of nervous systems: Mammals, 3,* 443–450). https://doi.org/10.1016/B0-12-370878-8/00081-1

Levy, R. (2024). The prefrontal cortex: From monkey to man. *Brain, 147*(3), 794–815. https://doi.org/10.1093/brain/awad389

Petrides, M., & Pandya, D. N. (1994). Comparative architectonic analysis of the human and the macaque frontal cortex. In F. Boller & J. Grafman (Eds.), *Handbook of neuropsychology: Vol. 9* (pp. 17–58). Elsevier.

Preuss, T. M. (2000). What's human about the human brain? In M. S. Gazzaniga (Ed.), *The new cognitive neurosciences* (2nd ed., pp. 1219–1234). MIT Press.

Rose, J. E., & Woolsey, C. N. (1948). The orbitofrontal cortex and its connections with the mediodorsal nucleus in rabbit, sheep and cat. *Research Publications – Association for Research in Nervous and Mental Disease, 27*(1), 210–232.

Schnider, A., Treyer, V., & Buck, A. (2005). The human orbitofrontal cortex monitors outcomes even when no reward is at stake. *Neuropsychologia, 43*(3), 316–323. https://doi.org/10.1016/j.neuropsychologia.2004.07.003

Seamans, J. K., Lapish, C. C., & Durstewitz, D. (2008). Comparing the prefrontal cortex of rats and primates: Insights from electrophysiology. *Neurotoxicity Research, 14*(2–3), 249–262. https://doi.org/10.1016/10.1007/BF03033814

Semendeferi, K., Armstrong, E., Schleicher, A., Zilles, K., & Van Hoesen, G. W. (1998). Limbic frontal cortex in hominoids: A comparative study of area 13. *American Journal of Physical Anthropology, 106*(2), 129–155. https://doi.org/10.1002/(SICI)1096-8644(199806)106:2<129::AID-AJPA3>3.0.CO;2-L

Sowell, E. R., Thompson, P. M., Holmes, C. J., Jernigan, T. L., & Toga, A. W. (1999). In vivo evidence for post-adolescent brain maturation in frontal and striatal regions. *Nature Neuroscience, 2*(10), 859–861. https://doi.org/10.1038/13154

Stuss, D. T., Benson, D. F., Clermont, R., Della Malva, C. L., Kaplan, E. F., & Weir, W. S. (1986). Language functioning after bilateral prefrontal leukotomy. *Brain and Language, 28*(1), 66–70. https://doi.org/10.1016/0093-934X(86)90091-X

Stuss, D. T., & Levine, B. (2002). Adult clinical neuropsychology: Lessons from studies of the frontal lobes. *Annual Review of Psychology, 53*, 401–433. https://doi.org/10.1146/annurev.psych.53.100901.135220

Uylings, H. B., & van Eden, C. G. (1990). Qualitative and quantitative comparison of the prefrontal cortex in rat and in primates, including humans. *Progress in Brain Research, 85*, 31–62. https://doi.org/10.1016/s0079-6123(08)62675-8

Uylings, H. B., Groenewegen, H. J., & Kolb, B. (2003). Do rats have a prefrontal cortex? *Behavioural Brain Research, 146*(1–2), 3–17. https://doi.org/10.1016/j.bbr.2003.09.028

4

• • • • • • •

Prefrontal Cortex Structure and Organisation

4.1 Structure

The human brain is divided into multiple regions and subregions based on cytoarchitecture and function. On a basic level, all brain regions, although they are involved in different processes and functions, are fundamentally the same in terms of anatomical structure. Every cortical region contains neurons and each of these neurons is connected to many others by synapses. Although there is some deviation from the rule, this basic principle applies throughout the whole human brain and the brains of all other animals.

It is evident from the literature that the functional roles of a particular brain region are dependent largely on the way in which the neurons that form it are connected through multiple neuronal networks, and the way in which the neurons themselves are physically organised. The most widely used classification of cortical regions based on cytoarchitectural differences between regions was produced by Brodmann (1909). Brodmann divided the human cortex into fifty-one distinct areas based on differences in cytoarchitectural features, such as how densely packed cells are (see Figure 4.1). Many regions described by Brodmann have since been described by their own individual complex structural organisations and functions.

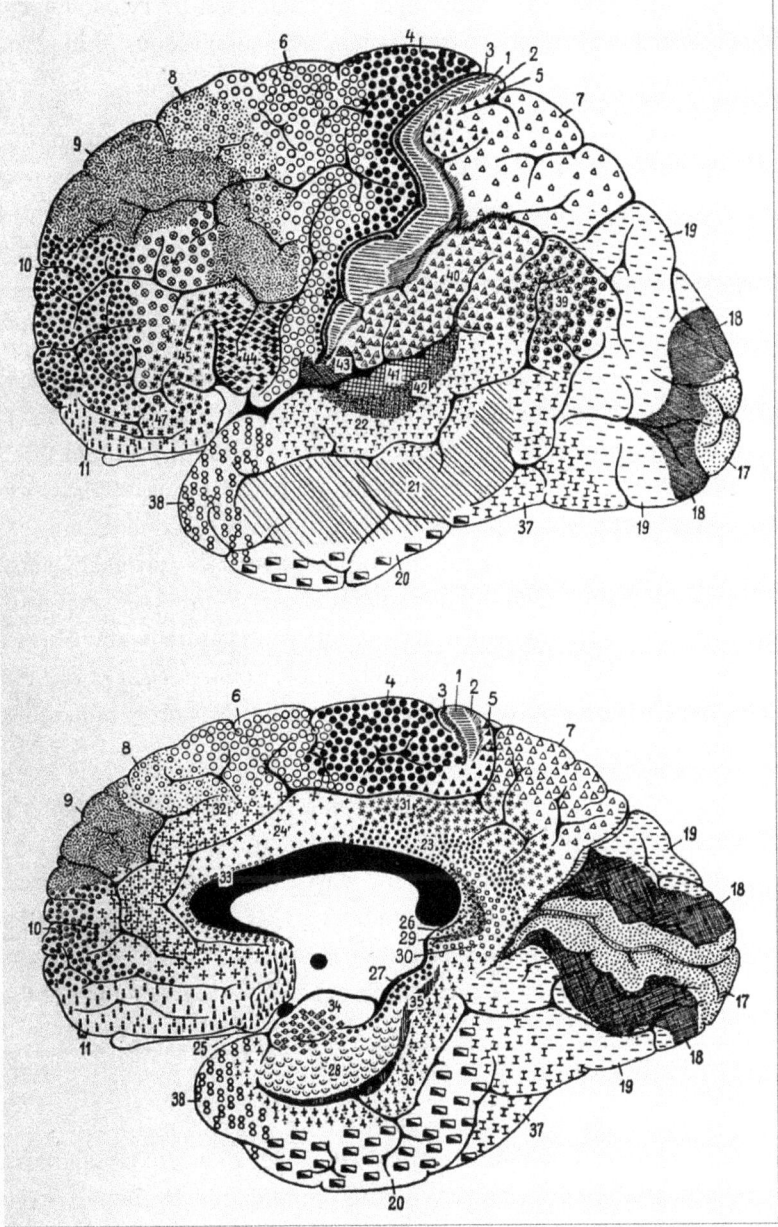

Figure 4.1 The original Brodmann's map, as published in Brodmann (1909).

PD-US-expired.

The prefrontal cortex (PFC), in mammals, including humans, much like other cortical regions, has traditionally been defined by cytoarchitectural features. It is thought that the cytoarchitectural subdivisions of the PFC in humans correspond to distinct functions.

In the somewhat simplified PFC of a rat, on which vast amounts of anatomical and physiological research are based, the PFC is typically divided into medial prefrontal cortex (occupying the medial wall of the region), orbital prefrontal cortex (dorsal to the olfactory bulb and rhinal sulcus) and agranular insular regions (anterior of the rhinal sulcus) (Van De Werd & Uylings, 2008).

Each of these regions is said to hold distinct functions, which correspond to a large extent to those found in primates and humans. Medial PFC is associated with rule switching, attention shifting (Furuyashiki & Gallagher, 2007), fear response (Morgan & LeDoux, 1995), working memory and behavioural flexibility (Heidbreder & Groenewegen, 2003). Orbital PFC has been found to be involved in reversal learning (McAlonan & Brown, 2003). Agranular insular cortex and orbitofrontal cortex as a whole are thought to have a role in the anticipation of reward and working memory (Kesner & Gilbert, 2007; Schoenbaum et al., 1998). However, there is no clear definition of unique functions belonging to each subdivision.

Prefrontal cortex architecture follows the same characteristic pattern of development, in terms of expansion, cell migration and lamination, as the rest of the cerebral cortex. So, in many ways, this region is no different to other areas of the brain. The basic architecture is established by genetics; however, further development of the PFC continues for over twenty years in humans, and how it develops can be influenced by a number of internal and external factors. Perhaps it is the fine-detailed changes in structure that continue for this extended time that result in such a highly complex cortical region.

The basic anatomical architecture of the PFC is not known to be significantly dissimilar to other brain regions. With this in mind, initially there is no clear anatomical reason why the connectivity of the PFC should be significantly different to that seen elsewhere in the cerebral cortex. However, it is clear that the PFC is involved in some of the most complex and evolutionary advanced cognitive processes; therefore, it is reasonable to suggest that something must be different in its anatomy for the PFC to be able to carry out such complex tasks.

4.2 Topographic Organisation

The organisation of cortical connections is highly complex. However, despite vast differences in function and connectivity between brain regions, there is a recurring theme in organisation throughout the brain. Topographic organisation has been described as a hallmark feature of cortical organisation among vertebrates and is widely as necessary for complex brain function. Topographic connectivity refers to point-to-point mappings, which preserve the spatial arrangements of information; nearby locations in a source region connect to nearby locations in a target region, maintaining the same spatial organisation. This type of organisation is found in connections and networks throughout the brain.

Topographic maps allow for efficient local computations by grouping together neurons that interact the most or are consistently activated at the same time. This reduces metabolic costs in terms of wiring and efficiently deals with the allocation of resources.

The existence of ordered representations of sensory processes has long been known and accepted among experts. The topographically ordered physiological and anatomical organisation has been described for multiple regions of cerebral cortex including motor, sensory, auditory, visual and temporal cortex – this type of organisation is very much considered standard across the brain.

Understanding anatomical organisation of a given brain region or structure is important for gaining a clear overall picture, including that of function. The anatomical connectivity of a given cortical region is often a strong indicator of the overlying physiological organisation. That is, how a region is connected anatomically can often provide us with important detailed information about how it works in terms of function. So anatomical connectivity is typically a sensible starting point. For instance, in somatosensory cortex, ordered representations correspond to the physical spatial layout of structures on the surface of the body (Figure 4.2).

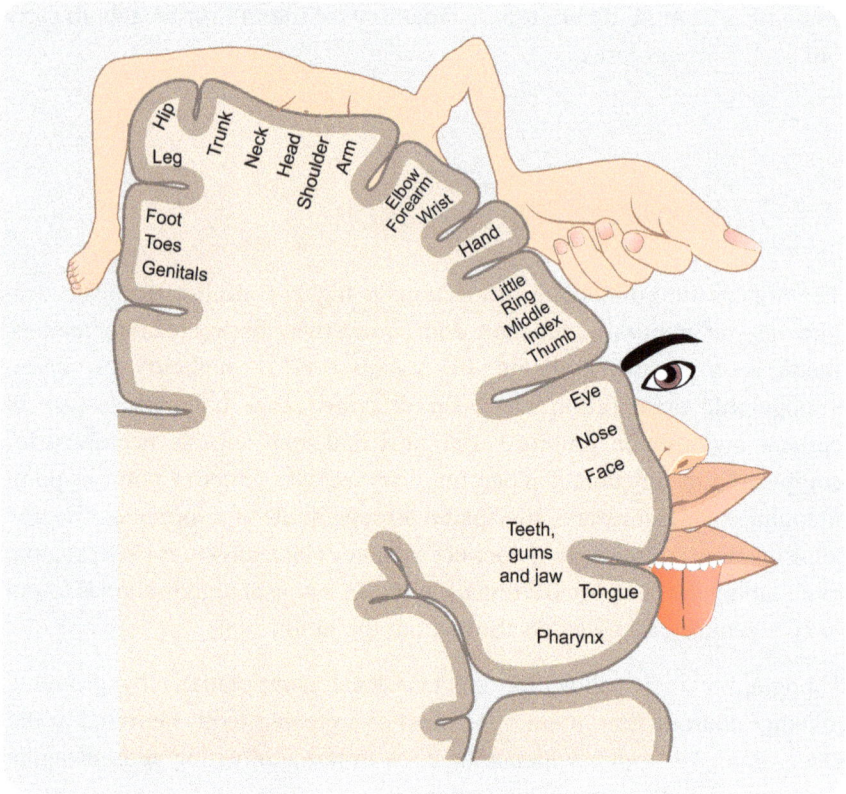

Figure 4.2 Sensory homunculus

Original work: OpenStax College / Derivative work: Ederporto https://creativecommons.org/licenses/by/3.0/.

The functional organisation of a given cortical region is thought to be highly dependent upon underlying structural organisation. Clear relationships between structure and function are common across biological systems, with structural organisation providing the basis and physiological constraints of function. It is suggested that the ability of a group of neurons to synchronise their activity depends on their ordered structural connectivity.

Based on evidence from other cortical regions, the identification of an ordered functional organisation within the PFC implies the existence of an underlying ordered arrangement of anatomical connections. To date there is no widely accepted organisational pattern of PFC connections, and there have been few detailed anatomical studies of the arrangement of PFC projections. To date there is no commonly accepted structural organisation of the PFC, particularly in terms of cortico-cortical connections. There have been few detailed and systematic anatomical studies of the organisation of its projections. The current literature has identified evidence for an ordered organisation of the PFC in terms of physiology; however, the precise circuitry is evidently more complex than the more clearly understood cortical regions (e.g. visual and motor cortex) and is yet to be fully defined. To clearly establish the nature of physiological organisation within the PFC, it is necessary to first gain a more detailed picture of the underlying anatomical connectivity. Some recent advances in neuroanatomy have begun to unravel the complex connectivity, pointing towards differences in comparison to other regions when studied in a finer scale than has been explored before (Bedwell & Tinsley, 2018).

The organisation of cortical connections, not just within the PFC but widespread across the brain, is undoubtedly highly complex and will not be fully understood for some time. One thing that is undisputed is that brain connectivity is not straightforward but is relatively similar across the brain. We know there can be vast differences in function and specific details of connectivity between brain regions, but there is a recurring theme in cortical organisation that can almost be applied as a rule. That rule is of topographic organisation. This point-to-point mapping of connections has been described as a hallmark feature of cortical organisation, at least in vertebrates (Thivierge & Marcus, 2007). Such ordered connectivity has even been considered as a required component for complex brain function (Kaas, 1997; Thivierge & Marcus, 2007), suggesting that if the brain was not organised in this highly ordered fashion, complex cognition we associated with high-order functions like decision-making would not be possible. Logically, topographic ordering of connections allows for efficient information transfer, meaning lower metabolic costs and allocation of resources (Kaas, 1997; Sereno & Huang, 2006).

Topographic connectivity refers to point-to-point mappings, which preserve the spatial arrangements of information; nearby locations in a source region connect to nearby locations in the target region, maintaining the same spatial organisation.

The existence of topographically mapped connections within the brain has been known for a long time, with some of the first observations in the early twentieth century (Penfield & Boldrey, 1937). Topographically ordered anatomical and physiological connectivity has been established in motor, sensory, auditory, visual and temporal cortices, in a variety of mammalian species – rodents, cats, non-human primates and humans (Albright & Desimone, 1987; Arcaro et al., 2009; Hafting et al., 2005; Lemon, 2008; Merzenich et al., 1975; Penfield & Boldrey, 1937; Welker, 1971; Woolsey, 1967). So, such organisation is quite confidently established as a general feature of the mammalian brain.

To date, there is no widely accepted organisational structure of the PFC, but anatomical studies in rodents (Bedwell et al., 2014, 2015, 2017; Bedwell & Tinsley, 2018) suggest similar ordering to other regions on a large scale (PFC subregions), but a distinctly different pattern of organisation when examined on a finer scale. This will be explored further later in this chapter.

4.3 Topographic Prefrontal Connections

The topographic organisation of the brain can be seen from multiple perspectives (anatomy, physiology, function) and is clearly prominent throughout brain regions and structures. This type of ordering is so prominent that for a long time it has been viewed as a fundamental element of cortical connectivity (Kaas, 1997). Many experts assumed this ordered arrangement is necessary for the complex processes the brain is capable of, without there being chaos. Even the highly complex PFC has been described as being topographically organised (Goldman-Rakic, 1988; Pandya et al., 1971), and indeed, modern evidence does still show this to be generally true (Bedwell et al., 2014, 2015, 2017; Bedwell & Tinsley, 2018). But this is not the whole picture. Although the PFC does appear at a superficial level to be structured in the same topographically ordered manner as what we understand to be the way of the rest of the cortex, more modern, fine-scale, evidence shows this is not the case when we look closer. Recent findings show us a different, more contradictory picture.

Advances in the investigation of anatomical connections in recent years have allowed for a vastly improved visualisation of cortical connections, in much greater resolution and accuracy than was previously possible. Modern connection studies using neuroanatomical tracers on a fine scale, do confirm previous ideas to a degree, that the PFC displays gross-level topologically organised connections. That is, magnified to a certain scale, it does appear as expected based on the known organisation of other regions. Evidence confirms that the PFC forms part of a large-scale, topographically arranged cortical pathway, and this has been replicated multiple times, in various species and exploring varied projections (Berendse et al., 1992; Olson & Musil, 1992; Schilman et al., 2008). Importantly though, the reported topographic ordering had only been described based on cytoarchitecturally distinct subregions, so on a fairly large scale.

4.4 Reciprocity and Alignment of Prefrontal Cortex

Historically, the connections of the PFC have been assumed to be reciprocal, largely because this is a common feature of cortical structure. There is indeed clear evidence for reciprocal connections in the PFC, but importantly, there is also growing evidence for non-reciprocal connections in this region. Early indications of a different structural pattern in the PFC came from reports of input and output connections missing each other in motor and sensory cortices (Flaherty & Graybiel, 1994). The result of what the authors described as missed connections is topographically mismatched pairs (non-reciprocal connectivity). Regions they studied did not receive input from the same regions elsewhere in the cortex that they were sending signals to. Such mismatching has also been observed in frontal cortices (McFarland & Haber, 2002). These findings began to provide a basis for the idea that cortical connections do not have to be reciprocal for optimum functioning, as had been assumed.

Bedwell and Tinsley (2018) provide evidence from fine-scale tract tracing studies in rats to show the anatomical connectivity of the PFC is quite different to expected, when visualised on a fine scale. That is, PFC connections are not always reciprocal. Findings from the studies clearly showed differential ordering of input and output connections in multiple PFC pathways. Further, there is evidence of an organisational gradient in the relationship between input and output connections from anterior to posterior – becoming increasingly differentiated closer to the anterior pole of the PFC. Investigations into the functional connectivity of the same region in humans, using event-related potential data (Katwa et al., in preparation) in decision-making, also suggest a gradient in the organisational pattern in terms of degree (how highly connected a region is).

DISCUSSION POINT

Is it more valuable to determine structural or functional organisation of the PFC?

4.5 Hierarchical Organisation

The PFC is called an association area, which means it is involved in lots of high-order processes. A high-order process comes at the top of what we call the processing hierarchy. A processing hierarchy describes the order in which different things work together, with the most important or most complicated things at the top of the pyramid (Figure 4.3). This makes the PFC a bit like the control centre or the CEO of the brain.

In a traditional model of hierarchical organisation, the PFC is positioned at the top of a processing hierarchy. In a hierarchical model, connections would travel from primary sensory cortex, followed by secondary sensory cortex and association areas, then reaching the top of the processing hierarchy, for example the PFC. This is followed by return connections travelling to secondary motor cortex followed by primary motor cortex; it is understood that reciprocal connections exist between source and target regions at each level of the hierarchy. Based on this understanding of cortical connectivity, it is thought that all cortical networks must contain a significant level of reciprocity in order to function, making it a fundamental structural component.

Traditional models of hierarchical organisation indicate that the PFC must be arranged in some kind of logical order to enable it to successfully connect with other cortical regions, as do historical theories of reciprocal

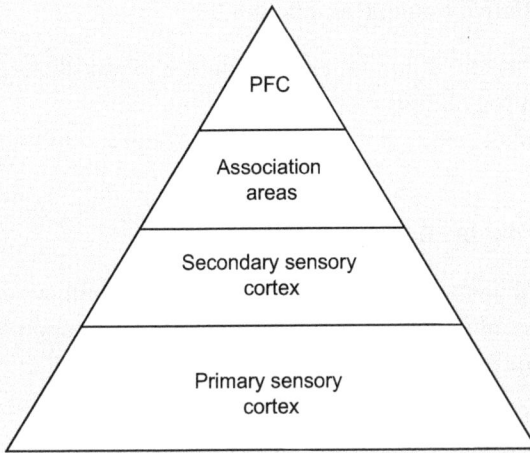

Figure 4.3
Processing hierarchy

connectivity. Hebb (1949) stated that an ordered map is guaranteed as long as the initial projection is ordered. If the initial input projection from a secondary sensory region is topographically ordered (which previous findings state they are), then according to Hebb's law PFC cortical connections would also have to be topographically ordered.

The organisation of the frontal lobe, specifically the PFC and how it contributes to decision-making, is very complex compared to some other brain regions we have a more complete understanding of. It has long been accepted by researchers that there is no simple localisation of decision-making, or even executive function as a whole. A decision is so complex, it involves multiple brain regions, both cortical and subcortical, and multiple networks.

Although there is a long way to go, knowledge does continue to grow and research has advanced over recent years. Evidence has started to reveal some interesting details of prefrontal structure that had been previously undefined. Models have been developed in an attempt to offer some logical explanation to the complexity of this brain region and its associated functioning, including the way decision-making is organised. A common emerging theme across many models is that of a hierarchy, or gradient, in the organisation of connections, whether that be anatomical, physiological, functional or behavioural.

To get a clear picture of the basic hierarchical organisation that can be seen, a telephone ringing is a simple example to follow:

1. Over time you have learned a simple behavioural response to your phone pinging, when it alerts you to a text message.
2. When you hear the familiar message tone, you pick up your phone and see the message.

So what's happening in your brain here?

The theoretical concept of hierarchical organisation of cognitive function in this way, although very simplified in this example (Figure 4.4), is supported with anatomical evidence (Bedwell & Tinsley, 2018).

4.6 Neuroplasticity

Although we consider the human brain to be fully developed by early adulthood, it is important to note that the brain never really stops changing. As we experience and learn new things each day, and form new memories, we continue to build and strengthen new connections between neurons. Other connections are weakened over time if they are not used. This is neuroplasticity.

4.7 Structural Plasticity

The brain's ability to create and build on connections based on experience is vital to our learning as we develop. Every time a child experiences something new, structural changes occur in the brain. The more something is experienced or practised, the stronger new connections become. Without this structural neuroplasticity cognitive development would not be possible.

4.8 Functional Plasticity

Functional neuroplasticity is beneficial in the case of brain damage. If specific areas of the brain are damaged, often connections elsewhere

At the top of the hierarchy is where responses are not constrained by episodic information or prior knowledge, or context. This is the level of control associated with the very frontal pole of the prefrontal cortex. Here it is thought we integrate information from sub-goals and specific conditions. The complex processing that takes place here allows you to consider decisions on a more general level, allowing for alternatives and weighing up possible outcomes.

Another level up in the hierarchy is where we consider cases where responding should not be constrained by context, or exceptions. At this level your brain uses your prior knowledge and experiences to guide your response. For instance, if you were aware of a family member having a surgical procedure that day, or if you are awaiting on an important response from someone, you might look at the text message in the lecture, regardless of the context.

At the next level of the hierarchy, your response might not be controlled by sensory input anymore, but contextual information. So, it considers the current context in how you respond to the message alter. For example, if you are driving or in the middle of a lecture, you might not pick up the phone to read the message, but if you are sat on the sofa watching TV, you likely will.

At the lowest level, premotor regions in your brain are guiding your behavioural response, to pick up the phone in response to a sensory stimulus. If the sensory stimulus (the ring tone) varies, your response might vary too. For example, if you set up your ringtones so you hear a different sound for family and work colleagues. You might be more inclined to wait longer to read one than the other.

PFC

Association areas

Secondary sensory cortex

Primary sensory cortex

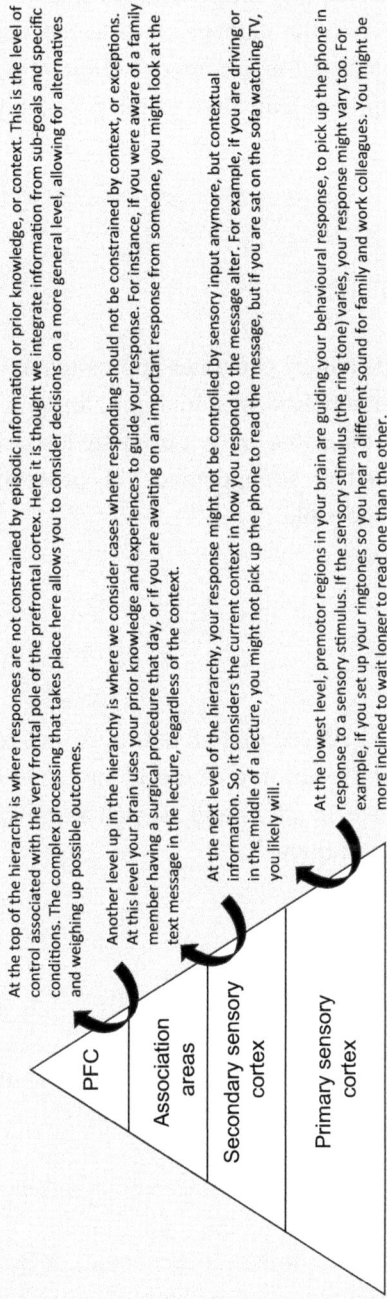

Figure 4.4 Processing hierarchy

can be strengthened and reorganised to compensate. A clear example of this is when people suffer a stroke, they are sometimes able to rehabilitate some abilities that were lost due to the damage, such as talking; this is because other networks are reorganising and compensating for the damaged network. The lost ability improves over time as the newly organised connections become stronger.

4.9 Synaptic Pruning

In the first few years of life a baby generates approximately 15,000 synapses per neuron. That number is halved by adulthood, through a process called synaptic pruning. Connections that are not used are lost, while connections that are frequently activated are strengthened. This process allows the brain to become as efficient as possible.

4.10 Neurogenesis

Until fairly recently it was believed that no new brain cells could be created after very early childhood. However, recent evidence shows we can actually generate new neuronal cells in adulthood, as well as creating new connections and pathways between them.

Key Points

- Structural organisation of the brain can be determined in multiple ways, including how cells are organised and how they are connected, both structurally and functionally.
- In terms of architecture, the PFC is organised no differently to other brain regions.
- In terms of topographic organisation of connections, at face value, the PFC is organised as expected, based on what we know of other cortical regions.

REFERENCES

Albright, T. D., & Desimone, R. (1987). Local precision of visuotopic organization in the middle temporal area (MT) of the macaque. *Experimental Brain Research*, *65*(3), 582–592. https://doi.org/10.1007/BF00235981

Arcaro, M. J., McMains, S. A., Singer, B. D., & Kastner, S. (2009). Retinotopic organization of human ventral visual cortex. *Journal of Neuroscience: The Official Journal of the Society for Neuroscience*, *29*(34), 10638–10652. https://doi.org/10.1523/JNEUROSCI.2807-09.2009

Bedwell, S. A., Billett, E. E., Crofts, J. J., MacDonald, D., & Tinsley, C. J. (2015). The topology of connections between rat prefrontal and temporal cortices. *Frontiers in Systems Neuroscience*, *9*, 80. https://doi.org/10.3389/fnsys.2015.00080

Bedwell, S. A., Billett, E. E., Crofts, J. J., & Tinsley, C. J. (2014). The topology of connections between rat prefrontal, motor and sensory cortices. *Frontiers in Systems Neuroscience*, *8*, 177. https://doi.org/10.3389/fnsys.2014.00177

Bedwell, S. A., Billett, E. E., Crofts, J. J., & Tinsley, C. J. (2017). Differences in anatomical connections across distinct areas in the rodent prefrontal cortex. *European Journal of Neuroscience*, *46*(3), 2005–2016. https://doi.org/10.1111/ejn.13521

Bedwell, S. A., & Tinsley, C. J. (2018). Mapping of fine scale rat prefrontal cortex connections: Evidence for detailed ordering of inputs and outputs connecting the temporal cortex and sensory-motor regions. *European Journal of Neuroscience*, *48*(3), 1944–1963. https://doi.org/10.1111/ejn.14068

Berendse, H. W., Galis-de Graaf, Y., & Groenewegen, H. J. (1992). Topographical organization and relationship with ventral striatal compartments of prefrontal corticostriatal projections in the rat. *Journal of Comparative Neurology*, *316*(3), 314–347. https://doi.org/10.1002/cne.903160305

Brodmann, K. (1909). *Beschreibung der einzelnen Hirnkarten IV. kapitel in Vergleichende Lokalisationslehre der Grosshirnrinde* [Description of individual brain maps. Chapter IV in *Localisation in the cerebral cortex*]. Verlag von Johann Ambrosias Barth.

Flaherty, A. W., & Graybiel, A. M. (1994). Input-output organization of the sensorimotor striatum in the squirrel monkey. *Journal of Neuroscience: The Official Journal of the Society for Neuroscience*, *14*(2), 599–610. https://doi.org/10.1523/JNEUROSCI.14-02-00599.1994

Furuyashiki, T., & Gallagher, M. (2007). Neural encoding in the orbitofrontal cortex related to goal-directed behavior. *Annals of the New York Academy of Sciences*, *1121*, 193–215. https://doi.org/10.1196/annals.1401.037

Goldman-Rakic, P. S. (1988). Topography of cognition: Parallel distributed networks in primate association cortex. *Annual Review of Neuroscience*, *11*, 137–156. https://doi.org/10.1146/annurev.ne.11.030188.001033

Hafting, T., Fyhn, M., Molden, S., Moser, M. B., & Moser, E. I. (2005). Microstructure of a spatial map in the entorhinal cortex. *Nature*, *436*(7052), 801–806. https://doi.org/10.1038/nature03721

Hebb, D. O. (1949). *The organization of behavior: A neuropsychological theory.* Wiley and Sons.

Heidbreder, C. A., & Groenewegen, H. J. (2003). The medial prefrontal cortex in the rat: Evidence for a dorso-ventral distinction based upon functional and anatomical characteristics. *Neuroscience and Biobehavioral Reviews, 27*(6), 555–579. https://doi.org/10.1016/j.neubiorev.2003.09.003

Kaas, J. H. (1997). Topographic maps are fundamental to sensory processing. *Brain Research Bulletin, 44*(2), 107–112. https://doi.org/10.1016/S0361-9230(97)00094-4

Katwa, G., Bedwell, S. A., Rogers, J., & Brookes, M. (in preparation). The functional network structure of lexical decision-making.

Kesner, R. P., & Gilbert, P. E. (2007). The role of the agranular insular cortex in anticipation of reward contrast. *Neurobiology of Learning and Memory, 88*(1), 82–86. https://doi.org/10.1016/j.nlm.2007.02.002

Lemon, R. N. (2008). An enduring map of the motor cortex. *Experimental Physiology, 93*(7), 798–802. https://doi.org/10.1113/expphysiol.2007.039081

McAlonan, K., & Brown, V. J. (2003). Orbital prefrontal cortex mediates reversal learning and not attentional set shifting in the rat. *Behavioural Brain Research, 146*, 97–103. https://doi.org/10.1016/j.bbr.2003.09.019

McFarland, N. R., & Haber, S. N. (2002). Thalamic relay nuclei of the basal ganglia form both reciprocal and nonreciprocal cortical connections, linking multiple frontal cortical areas. *Journal of Neuroscience, 22*(18), 8117–8132. https://doi.org/10.1523/JNEUROSCI.22-18-08117.2002

Merzenich, M. M., Knight, P. L., & Roth, G. L. (1975). Representation of cochlea within primary auditory cortex in the cat. *Journal of Neurophysiology, 38*(2), 231–249. https://doi.org/10.1152/jn.1975.38.2.231

Morgan, M. A., & LeDoux, J. E. (1995). Differential contribution of dorsal and ventral medial prefrontal cortex to the acquisition and extinction of conditioned fear in rats. *Behavioral Neuroscience, 109*(4), 681–688. https://doi.org/10.1037/0735-7044.109.4.681

Olson, C. R., & Musil, S. Y. (1992). Topographic organization of cortical and subcortical projections to posterior cingulate cortex in the cat: Evidence for somatic, ocular, and complex subregions. *Journal of Comparative Neurology, 324*(2), 237–260. https://doi.org/10.1002/cne.903240207

Pandya, D. N., Dye, P., & Butters, N. (1971). Efferent cortico-cortical projections of the prefrontal cortex in the rhesus monkey. *Brain Research, 31*(1), 35–46. https://doi.org/10.1016/0006-8993(71)90632-9

Penfield, W., & Boldrey, E. (1937). Somatic and sensory representation in the cerebral cortex of man as studied by electrical stimulation. *Brain, 60*(4), 389–443. https://doi.org/10.1093/brain/60.4.389

Schilman, E. A., Uylings, H. B., Galis-de Graaf, Y., Joel, D., & Groenewegen, H. J. (2008). The orbital cortex in rats topographically projects to central parts of the caudate-putamen complex. *Neuroscience Letters, 432*(1), 40–45. https://doi.org/10.1016/j.neulet.2007.12.024

Schoenbaum, G., Chiba, A., & Gallagher, M. (1998). Orbitofrontal cortex and baso-lateral amygdala encode expected outcomes during learning. *Nature Neuroscience, 1*, 155–159. https://doi.org/10.1038/407

Sereno, M. I., & Huang, R. S. (2006). A human parietal face area contains aligned head-centered visual and tactile maps. *Nature Neuroscience, 9*(10), 1337–1343. https://doi.org/10.1038/nn1777

Thivierge, J. P., & Marcus, G. F. (2007). The topographic brain: From neural con-nectivity to cognition. *Trends in Neurosciences, 30*(6), 251–259. https://doi.org/10.1016/j.tins.2007.04.004

Van De Werd, H. J., & Uylings, H. B. (2008). The rat orbital and agranular insular prefrontal cortical areas: A cytoarchitectonic and chemoarchitectonic study. *Brain Structure and Function, 212*(5), 387–401. https://doi.org/10.1007/s00429-007-0164-y

Welker, C. (1971). Microelectrode delineation of fine grain somatotopic organization of (SmI) cerebral neocortex in albino rat. *Brain Research, 26*(2), 259–275. https://doi.org/10.1016/S0006-8993(71)80004-5

Woolsey, T. A. (1967). Somatosensory, auditory and visual cortical areas of the mouse. *The Johns Hopkins Medical Journal, 121*(2), 91–112.

5

· · · · · · ·

Neurotransmitters
and Neurophysiology

Communication and information transfer in the brain depends on both chemical and electrical signalling between neurons. Neurotransmitters are vital chemicals that are released by neurons through the process of synaptic transmission. This process allows neurons to communicate with one another, and for information to be transported all over the brain. Because of their importance in information transfer, neurotransmitters are very important in cognitive functions. There are a wide variety of neurotransmitters involved in cognition, with a wide range of roles. Many aspects of neurotransmission are influenced by the type of neurotransmitter involved.

There are three types of chemical structure when it comes to neurotransmitters. These are monoamines, amino acids and peptides. Monoamines include dopamine, noradrenaline, adrenaline and serotonin. Amino acids include GABA and glutamate. Peptides include endorphins and oxytocin. For the purpose of understanding decision-making, we are mostly interested in monoamines.

5.1 Excitation and Inhibition

Excitatory refers to neurotransmitters that produce an excitatory effect on a neuron. This could be compared to putting your foot down on the accelerator when driving a car. On the other hand, inhibitory neurotransmitters produce an inhibitory effect. This is like putting your foot on the brake in the car, to have a slowing down or dampening effect. For the complex functions we carry out every day, it is important that the brain is able to both increase and decrease activity in this way, and to do so efficiently. For instance, if inhibitory neurotransmitters were not available in the brain, the system could become overwhelmed and get a bit out of control.

5.2 Systems

To understand the underpinnings of decision-making, the aspects of neurotransmission we are really interested in are the neurotransmitter systems. Neurotransmitters that are organised into systems are produced by neurons whose cell bodies are located subcortically – below the cortex and in the brainstem. The axons from these subcortical neurons project widely, through to the whole cortex, so they reach all areas of the brain.

Each of these neurotransmitters is released by a different set of neurons that together form a neurotransmitter system. For example, one of these systems is the dopaminergic system. All neurons in the dopaminergic system produce a neurotransmitter called dopamine. We sometimes call these dopaminergic neurons because these neurotransmitter systems reach all areas of the cortex with their projections. Each system can affect a large variety of behaviours and cognitive processes. Some even overlap between systems, in that multiple neurotransmitter systems affect one behaviour. An example of this is decision-making.

Dopaminergic Cells and Parkinson's

Degeneration of the brain's dopaminergic neurons is linked to the onset of Parkinson's disease (Zhou et al., 2023). Dopaminergic neurons are particularly vulnerable to damage because of their extensive connectivity across the brain.

Serotonin is the neurotransmitter released by the serotonergic system. The cell bodies of the serotonergic system are found in several clusters around the brain, including in the midbrain, the pons and the medulla (Hornung, 2010).

The neurons in the serotonergic system project to the hypothalamus, hippocampus and amygdala. These are all part of the limbic system, and serotonin is suggested to play a neuromodulatory role here (Hensler, 2006). They also project to other areas including the striatum, cortical areas, the cerebellum and the thalamus (Nair et. al., 2020; Oostland & Van Hooft, 2013; Wai et al., 2011). This system is very widely connected! As a result of being so widely connected to so many areas of the brain, the serotonergic system influences a very wide variety of behaviours, including but not limited to our sleep, mood, sexual behaviour, eating and memory (Bacqué-Cazenave et al., 2020; Brown et al., 2012; Olivier et al., 2019; Steiger, 2004). Many of these behaviours are regulatory and help us meet our basic needs, but some are important in higher-order cognitive processes like decision-making. For example, memory is a vital component of decision-making. Serotonin has also been linked to mood states – perhaps most commonly known, depression (Coppen, 1967). Mood and depression can influence how we make decisions (Gear et al., 2017). Some of the most widely prescribed drugs used to treat depression are serotonin-specific reuptake inhibitors (SSRIs). These drugs work at alleviating the symptoms of depression by acting directly on the serotonergic system. They increase the amount of serotonin in the synaptic cleft by inhibiting its presynaptic uptake. Prozac is a well-known brand name of a widely prescribed SSRI. While SSRI treatment for depression is shown to be beneficial, because the serotonergic system is important in such a wide range of behaviours and processes, taking SSRIs can also have a wide range of undesirable effects as well. Examples are effects on sleep and appetite. The serotonin hypothesis of depression is over fifty years old; however, more recently, evidence has suggested there may not be a direct link between low serotonin and depression at all (Moncrieff et al., 2023).

We know that serotonin is strongly linked to memory, and specifically the creation of new memories (Cedar & Schwartz, 1972). For example, research has established a link between lower levels of serotonin and difficulties in forming new memories (Coray & Quednow, 2022). A diminished ability to form new memories can have a significant impact on decision-making – because memories, prior learning and existing knowledge we have of a given stimulus or circumstance play a vital role in the decisions we make.

People who take SSRIs long term often show deficits in long-term memory, therefore having an undesirable effect on other areas of cognition, including decision-making.

Noradrenaline (interchangeable with norepinephrine) is a neurotransmitter emitted by cells in the noradrenergic system. This neurotransmitter system originates in the locus coeruleus and, like the serotonergic system, is known to project to multiple regions and structures – including the thalamus and hypothalamus as well as cortical regions. Perhaps one of the most notable projections of this system, whilst we are considering decision-making, is the prefrontal cortex. As established in earlier chapters, the prefrontal cortex is vital when it comes to coordinating decisions, so it can therefore be assumed that the noradrenergic system is involved in decision-making in some capacity.

Noradrenaline also plays a role in sleep and has a really interesting part here. The receptors in the thalamus effectively put the brain into sleep mode, then noradrenergic cells shut off during REM sleep. The only real difference between being awake and asleep is noradrenaline (Gottesmann, 2011), so it is actually quite an important chemical for our day-to-day functioning. Attention also plays an important role in decision-making. Some researchers have suggested that a causal factor of attention deficit hyperactivity disorder may be disruption in noradrenalin levels (Biederman & Spencer, 1999).

Dopamine is the neurotransmitter of the dopaminergic system. In comparison to the serotonergic and noradrenergic systems, this one is more complex. The dopaminergic system has three subsystems within itself. These are the nigrostriatal, mesolimbic and mesocortical systems. The three subsystems can be most easily differentiated by their origin – as in the anatomical location where the neuron cell bodies involved in the subsystem can be found.

For the purpose of understanding decision-making, here we consider the dopaminergic system as a whole. There are many types of dopamine receptors, but they can all be categorised into two types – they are either located on the postsynaptic dendrite or the presynaptic terminal.

Antipsychotic drugs have been valuable in aiding the understanding of the complex dopaminergic system. Many antipsychotic drugs work as D2 antagonists. For example, a common antipsychotic drug, chlorpromazine, blocks this specific type of dopamine receptors. The effect is to stop dopamine being transported via the brain's networks, reducing the amount that is available in the cells. These types of antipsychotic drugs are very effective at reducing the positive symptoms of schizophrenia, such as hallucinations and delusions. This shows us that dopamine must be involved in their manifestation. Without the same amount of dopamine available, the symptoms reduce.

The dopaminergic system is known to be significantly involved in working memory, novelty seeking and attention (Alcaro et al., 2007; Carr et al., 2017). All of these contribute to decision-making, making this another neurotransmitter system that is important in the decision process.

DISCUSSION POINT

How could neurotransmitters contribute to deficits in decision-making, such as seen in schizophrenia?

5.3 The Neurochemical Process of Decision-Making

5.3.1 The Synapse

Neurons do not physically touch each other, but it is vital that they communicate and transfer information. Across the central nervous system, where an axon terminal comes close to a dendrite of another neuron, there is a small gap between the two. Across this gap, a synapse is formed. It is through the synapse that the two neurons communicate, and information is transmitted around the brain and to the rest of the body. This information transfer is, however, much more sophisticated than an electrical signal transmitting across the gap and continuing to the next neuron. When an electrical impulse reaches the axon terminal of the presynaptic neuron, it triggers the release of neurotransmitters, which carry the signal across the synapse to the postsynaptic neuron (Figure 5.1).

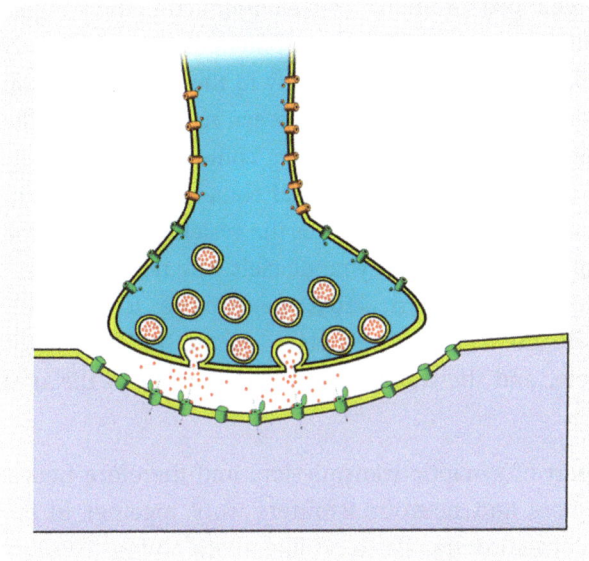

Figure 5.1 Synaptic transmission

Original work: Sheldahl / Derivative work: Dr Stacey Bedwell https:// creativecommons.org/ licenses/by-sa/4.0/.

To make things more complex, the neurotransmitters cannot be stored in the axon terminal and transported around networks freely. The axon terminal contains small holding vessels, called vesicles. The vesicles house the neurotransmitters.

Neurotransmitters are chemical messengers. Their role is to transmit signals from one neuron to the next. A single neuron can receive information in this way from many axon terminals. Similarly, a single neuron can form synapses with many neurons.

5.4 Electrical and Chemical Signalling

Synaptic transmission is the physiological process by which the pre-synaptic neuron communicates with the post-synaptic neuron. Information is passed from the cell body, down the axon of the pre-synaptic neuron as an electrical impulse. This part of the communication process is known as an action potential. When the electrical action potential reaches the axon terminal, it needs to be changed for information transfer across the synapse to be possible. The electrical impulse of the action potential, although highly efficient for getting information from the cell body to the end of the axon (which can be quite a distance!), cannot cross between neurons, across the synaptic cleft. This is where chemical messengers come in. The synaptic vesicles located in the pre-synaptic axon terminal contain specific neurotransmitters. When the electrical impulse reaches the vesicles, it triggers the release of the neurotransmitters into the synaptic cleft. The neurotransmitters carry the signal in chemical form across the synapse, to the post-synaptic dendrite, where they bind to receptors. At this point synaptic transmission is complete, and the message has been passed to the next neuron in the network.

Receptors are a vital part of synaptic transmission, and therefore neural communication. Receptors and neurotransmitters work together in the process of chemical information transfer a bit like a lock and key system. Just as it takes the correct key to be able to open a specific lock, and enter a room, it takes a specific neurotransmitter to fit the correct receptor and trigger changes in the post-synaptic cell. If a neuron does not possess the correct receptor for a given neurotransmitter, the neurotransmitter cannot bind to the neuron and will therefore not trigger any changes, and information transfer will not continue to that post-synaptic cell. Many drugs, both clinical and recreational, work via this same mechanism. A drug that is designed to bind to the same receptor as a specific neurotransmitter will block the receptor, meaning the neurotransmitter has nowhere to bind to and the effects of that neurotransmitter are therefore reduced. Essentially, these types of drugs block the action of the targeted neurotransmitter.

5.5 What Happens in Your Brain When You Make a Decision?

Researchers across neuroscience and psychology, as well as philosophers, have been attempting to explain the human decision-making process for a very long time. Although our understanding of decision-making has improved greatly, from many perspectives, over recent decades, we still do not have a clear answer to exactly how we make a decision and what influences it.

Different disciplines take different approaches to answering the overarching question of how we make a decision. In economics, for instance, research tends to focus on how we come to a correct or incorrect decision. Others may argue there is no such thing as a correct decision, as it is so subjective. In contrast, neuropsychology focuses on the actions of the brain and how it can compute such a complex task as a decision. As humans, we make thousands of decisions every day. Many decisions are very simple and not considered complex, a lot of which we may not even be conscious of. Other decisions are more complex and even considered life-changing in some instances. Despite the vast differences in decisions being made, they all have one thing in common – they need to be processed. Epstein (2003) devised the cognitive experiential theory of decision-making in an attempt to explain these complexities. The theory suggests there exists two distinct systems of decision-making, and these two systems differ in the way in which they process information for decisions to be made.

5.5.1 The Rational System

The rational system includes a type of decision processing that utilises what Epstein described as a free emotional perspective. This results in a more analytic and effortful decision-making process than the experiential system. Rational decisions are always conscious. This is the more reflective of the two systems. This system is said to involve structures including the hippocampus, insula and prefrontal cortex. We know from previous chapters that the prefrontal cortex is vital in the integration of memory recall and imagination of possible outcomes when considering consequences of a decision. Interestingly, patients with damage to specific parts of the prefrontal cortex consistently show impairments in rational decision-making (Manes et al., 2002). It is important to note, however, that the normal functioning of the prefrontal cortex, and therefore normal functioning in terms of rational decision-making, is not just relying on this region. The prefrontal cortex depends on other neural systems to function normally itself. Structures including the insula and hippocampus, and even other parts of the prefrontal cortex, need to be functioning normally for this complex area to contribute to the decision process. This neural overlapping is very important to the decision-making process and suggests a functional interconnection between memory, emotions and the cognitive processing of decisions.

5.5.2 The Experiential System

The experiential system is based more on emotions and is said to be more effortless, so more like emotive responses in this sense. Experiential decisions can be unconscious. This is the more impulsive of the two systems. Brain structures said to be involved in the experiential system include the amygdala and striatum. Both of these structures are known to be involved in fast responses to stimuli. The amygdala in particular is established as being involved in giving emotional value to stimuli (Davis & Whalen, 2001; Inman et al., 2020; LeDoux, 2017). This offers some explanation as to why it would be involved in decisions involving emotion.

5.6 Controlled versus Automatic Processes in Decision-Making

The 1980s saw the development of two important theories of executive function. The first (Shallice, 1982) suggested a two-component system that influences our choice of behaviour, or decision.

5.6.1 Contention Scheduling

Contention scheduling is a cognitive system that enables relatively automatic processing of information. This means it happens without conscious input, and therefore little cognitive effort. The theory suggests that the automated process has been learned over time, through a process of specific stimuli becoming cognitively linked with specific actions or routines in behaviour. Groups of behaviours or routines can also be linked together, to form a *schema*. A good example of this type of schema is when you see a red traffic light while driving. The external stimulus here is the red light, and the presence of the red light causes a series of automatic actions. For example, taking your foot off the accelerator, applying your foot to the brake pedal, determining how hard to press the brake pedal to stop in the right place, and so on. Once an action is begun by this system, it continues until it is inhibited by an incompatible process.

5.6.2 Supervisory Attentional System

The second of Shallice's systems is said to be required to effortfully direct attention through decision processes. This system, unlike contention scheduling, is not automatic, but consciously controlled. The theory states that the supervisory attentional system is only active in certain situations. These situations occur when there is no pre-existing schema for a behaviour, so the contention scheduling system would be unable to act. So, in other words, the supervisory attentional system applies in novel situations. Shallice theorised that damage to the frontal lobe would effectively disable the supervisory attentional system and leave all behaviours and actions to be controlled by contention scheduling. This links to our knowledge surrounding decision-making deficit in novel situations after frontal lobe damage (Godefroy & Rousseaux, 1997). This theory implies that patients with frontal lobe damage should show no deficits when it comes to everyday actions and decisions like stopping at a red light, but would not be able to make an appropriate response in a novel situation.

Following from Shallice (1982), Stuss and Benson (1986) theorised a model of hierarchical structure. Like Shallice, Stuss and Benson recognised the importance of the frontal lobes in regulating behaviour in normal situations.

Key Points

- The serotoninergic system is important to decision-making due to its involvement in memory, specifically the creation of new memories.
- The noradrenergic system is important to decision-making due to strong connections to the prefrontal cortex and its role in sleep.
- The dopaminergic system is important to decision-making due to its significant role in working memory.

REFERENCES

Alcaro, A., Huber, R., & Panksepp, J. (2007). Behavioral functions of the mesolimbic dopaminergic system: An affective neuroethological perspective. *Brain Research Reviews*, *56*(2), 283–321. https://doi.org/10.1016/j.brainresrev.2007.07.014

Bacqué-Cazenave, J., Bharatiya, R., Barrière, G., Delbecque, J.-P., Bouguiyoud, N., Di Giovanni, G., Cattaert, D., & De Deurwaerdère, P. (2020). Serotonin in animal cognition and behavior. *International Journal of Molecular Sciences*, *21*(5), 1649. https://doi.org/10.3390/ijms21051649

Biederman, J., & Spencer, T. (1999). Attention-deficit/hyperactivity disorder (ADHD) as a noradrenergic disorder. *Biological Psychiatry*, *46*(9), 1234–1242. https://doi.org/10.1016/s0006-3223(99)00192-4

Brown, G. R., & Hariri, A. R. (2012). The reciprocal interaction between serotonin and social behaviour. *Neuroscience & Biobehavioral Reviews*, *36*(2), 786–798. https://doi.org/10.1016/j.neubiorev.2011.12.009

Carr, G. V., Maltese, F., Sibley, D. R., Weinberger, D. R., & Papaleo, F. (2017). The dopamine D5 receptor is involved in working memory. *Frontiers in Pharmacology*, *8*, 666. https://doi.org/10.3389/fphar.2017.00666

Cedar, H., & Schwartz, J. H. (1972). Cyclic adenosine monophosphate in the nervous system of Aplysia californica: II. Effect of serotonin and dopamine. *Journal of General Physiology*, *60*(5), 570–587. https://doi.org/10.1085/jgp.60.5.570

Coppen, A. (1967). The biochemistry of affective disorders. *British Journal of Psychiatry*, *113*, 1237–1264.

Coray, R., & Quednow, B. B. (2022). The role of serotonin in declarative memory: A systematic review of animal and human research. *Neuroscience and Biobehavioral Reviews*, *139*. https://doi.org/10.1016/j.neubiorev.2022.104729

Davis, M., & Whalen, P. J. (2001). The amygdala: Vigilance and emotion. *Molecular Psychiatry*, *6*(1), 13–34. https://doi.org/10.1038/sj.mp.4000812

Epstein, S. (2003). Cognitive-experiential self-theory of personality. In T. Millon & M. J. Lerner (Eds.), *Handbook of psychology: Vol. 5. Personality and social psychology* (pp. 159–184). John Wiley & Sons, Inc.

Gear, T., Shi, H., Davies, B. J., & Fets, N. A. (2017). The impact of mood on decision-making process. *EuroMed Journal of Business*, *12*(3), 242–257. https://doi.org/10.1108/EMJB-04-2016-0013

Godefroy, O., & Rousseaux, M. (1997). Novel decision making in patients with prefrontal or posterior brain damage. *Neurology*, *49*(3), 695–701. https://doi.org/10.1212/wnl.49.3.695

Gottesmann, C. (2011). The involvement of noradrenaline in rapid eye movement sleep mentation. *Frontiers in Neurology*, *2*, 81. https://doi.org/10.3389/fneur.2011.00081

Hensler, J. G. (2006). Serotonergic modulation of the limbic system. *Neuroscience and Biobehavioral Reviews*, *30*(2), 203–214. https://doi.org/10.1016/j.neubiorev.2005.06.007

Hornung, J. (2010). The neuroanatomy of the serotonergic system. In C. P. Müller & B. L. Jacobs (Eds.), *Handbook of behavioral neuroscience* (Vol. 21, pp. 51–64). Elsevier.

Inman, C. S., Bijanki, K. R., Bass, D. I., Gross, R. E., Hamann, S., & Willie, J. T. (2020). Human amygdala stimulation effects on emotion physiology and emotional experience. *Neuropsychologia,* *145,* 106722. https://doi.org/10.1016/j.neuropsychologia.2018.03.019

LeDoux, J. E. (2017). Semantics, surplus meaning, and the science of fear. *Trends in Cognitive Sciences, 21*(5), 303–306. https://doi.org/10.1016/j.tics.2017.02.004

Manes, F., Sahakian, B., Clark, L., Rogers, R., Antoun, N., Aitken, M., & Robbins, T. (2002). Decision-making processes following damage to the prefrontal cortex. *Brain, 125*(3), 624–639. https://doi.org/10.1093/brain/awf049

Moncrieff, J., Cooper, R. E., Stockmann, T., Amendola, S., Hengartner, M. P., & Horowitz, M. A. (2023). The serotonin theory of depression: A systematic umbrella review of the evidence. *Molecular Psychiatry, 28,* 3243–3256. https://doi.org/10.1038/s41380-022-01661-0

Nair, S. G., Estabrook, M. M. & Neumaier, J. F. (2020). Serotonin regulation of striatal function. In C. P. Müller & K. A. Cunningham (Eds.), *Handbook of behavioural neuroscience* (Vol. 31, pp. 321–335). Elsevier. https://doi.org/10.1016/B978-0-444-64125-0.00018-9

Olivier, J. D. A., Esquivel-Franco, D. C., Waldinger, M. D., & Olivier, B. (2019). Serotonin and sexual behavior. In M. D. Tricklebank & E. Dal (Eds.), *The serotonin system: History, neuropharmacology, and pathology* (pp. 117–132). Academic Press.

Oostland, M., & Van Hooft, J. A. (2013). The role of serotonin in cerebellar development. *Neuroscience, 248,* 201–212. https://doi.org/10.1016/j.neuroscience.2013.05.029

Shallice, T. (1982). Specific impairments of planning. *Philosophical Transactions of the Royal Society of London. Series B, Biological Sciences, 298*(1089), 199–209. https://doi.org/10.1098/rstb.1982.0082

Steiger, H. (2004). Eating disorders and the serotonin connection: State, trait, and developmental effects. *Journal of Psychiatry & Neuroscience, 29*(1), 20–29.

Stuss, D. T., & Benson, D. F. (1986). *The frontal lobes.* Raven Press.

Wai, M. S., Lorke, D. E., Kwong, W. H., Zhang, L., & Yew, D. T. (2011). Profiles of serotonin receptors in the developing human thalamus. *Psychiatry Research, 185*(1–2), 238–242. https://doi.org/10.1016/j.psychres.2010.05.003

Zhou, Z. D., Yi, L. X., Wang, D. Q., Lim, T. M., & Tan, E. K. (2023). Role of dopamine in the pathophysiology of Parkinson's disease. *Translational Neurodegeneration, 12,* 44. https://doi.org/10.1186/s40035-023-00378-6

6

· · · · · · ·

Memory

Memory is a vital element of human brain function. Not only for the complex high-order processes like decision-making but also for many more, simpler and less cognitively demanding behaviours and actions we complete every day. For instance, your ability to read this text, to hold the book, to recall what each word means – these all require memory. To fully appreciate the importance of memory in executive functions of decision-making, one must have a grasp on the complexities of memory itself.

Before we begin to explore memory in any detail and consider how it applies in decision-making, it is important we first establish what we actually mean when we refer to memory. The term memory can refer to many different things, depending on the context. Generally, memory is defined by an ability to retain learned information and knowledge of past events or stimuli. Memory also involves retrieval – you may argue there is no value in retention of information if we cannot later retrieve and utilise it. So memory also refers to the retrieval of previously learned information. For memory to function, it must have the ability to logically organise retained information in some coherent way, that can be later accessed at will – and stored again.

The term memory has differentiation in its meaning depending on the perspective. For instance, in everyday society memory, or to say you remember, usually refers to bringing past events to mind (retrieving). For instance, you might say to a friend that you remembered to bring a pen to class, or you might recount what you did at the weekend. However, to a memory scientist, specifically a neuroscientist, memory is a function of experience based on anatomical and physiological networks in the brain. In psychology memory might mean the storage and retrieval of information, as well as the flow of information through some complex system. In the interest of investigating the cognitive neuroscience of decision-making, from a neuropsychological perspective here we are interested in memory as a neurobiological system and how that system allows for the storage and retrieval of information for complex decisions to take place.

Human memory can be considered a complex storage system, like a filing cabinet with information organised into different drawers and files according to specific categories. The information within the storage system can be stored according to a number of different criteria. There are several models of decision-making that attempt to explain and simplify the complexities of human decision-making using varied criteria for storage. Generally, the information stored within memory may be organised by content or by duration. If stored by duration, this refers to the length of time the information needs to be stored for and when it will need to be accessed again. Content refers to the type of information. Memory information may also be stored by capacity; this means how much information we need to store. Memory information can be further categorised by the way it is encoded. This means the way in which information is stored, or the cognitive processing involved.

The final categorisation is retrieval; this is how the information is accessed when we need it again – or when we recall it. All of these different ways of storing or categorising memory information is referred to as fractation.

6.1 Fractation of Memory

Duration of retention is one of the most commonly referred to fractations of memory. This appears in several models of memory. Based on duration, there are three specific types of memory – sensory, short-term and long-term. Sensory memory has the shortest duration and long-term memory the longest.

6.1.1 Sensory Memory

Within sensory memory information is retained for a maximum of a few seconds, a very short time. Sensory memory corresponds only to the here and now, what you are currently perceiving. Some information from sensory memory is transferred to short-term memory.

6.1.2 Short-Term Memory

Short-term memory is longer in duration than sensory memory; information here is retained from seconds to minutes. This type of memory refers to the brief period of time when you can recall something you just experienced. It is thought that short-term memory has a capacity of about seven items (Miller, 1956) – although this is debated.

6.1.3 Long-Term Memory

Information retained in long-term memory can be stored for multiple decades and still be easily recalled. The success of recall from long-term memory, however, can depend on how frequently the information is recalled and the depth of processing.

6.2 Classification by Information Type

Memory can alternatively be classified according to information type. Information stored as memories can be either explicit or implicit. Explicit, or declarative memory, is a type of memory that requires conscious recall and can be further divided into semantic, episodic and autobiographical memory.

- Semantic memories are independent of a context. In other words, information that has no relationship with a specific location, person, time or emotion. An example of a semantic memory is remembering that London is the capital of England. Semantic memories are important to us as humans because they allow the encoding of abstract knowledge; high-order processes like planning and decision-making require this.

- Episodic memories are specific to a context. These are more personal memories, related to emotions or personal associations. An example of an episodic memory could be recalling what your first home looked like.
- Autobiographical memories relate to specific events in your life. For example, remembering what happened on your tenth birthday.

The second classification of memory when organised by information type is procedural memory. Procedural memory is based on implicit learning. This includes things like motor skills, that follow a specific procedure or order of learnt movements. You might be engaging your procedural memory as you are working through this chapter. For instance, if you are making notes as you read the text, you are using your procedural memories of how to hold your pen and how to move it across the page to form letters and words. It is unlikely that as you do this you are consciously recalling each step of the writing process, as you form each letter.

6.3 Classification by Temporal Direction

To organise memory by temporal direction means to classify according to whether information is recalled from the past or is referring to something to be carried out in the future. This is a timeline-based form of classification that falls into two categories. Retrospective memory is to recall or remember information or events from the past. In contrast, prospective memory is to retain a memory for future intentions; for example, remembering to buy milk on the way home. A prospective memory can be either time based, like remembering a lecture at 9 a.m., or it can be event based and triggered by a cue, such as remembering you have a letter to post when you see a postbox.

DISCUSSION POINT

Identify the possible limitations to our understanding of memory that categorising it could pose.

6.4 Memory as a Cognitive Function

Developing explanations for how memories are stored is useful, but to really gain an insight into how memory informs decision-making one must consider it as a dynamic cognitive function. There are three main stages of processing that information goes through in the memory system. First, information is encoded into a suitable format, then stored, and later retrieved. This is a simplified explanation of memory structure but forms the basis of all models and theories of memory.

6.5 Encoding and Storage

The first stage of memory function, encoding, is where stimuli experienced within our environment need to be changed into a format that our memory systems can store and is also meaningful when later retrieved. Information may be encoded differently depending on the nature of the stimulus. If it is something we have seen, it is visual information. If something we have heard, it is acoustic information. If it involves meaning or understanding, this is semantic.

Have a go at testing your own encoding. Take a moment to read this phone number, then cover it over and try to recall it.

07786453389

Could you remember the whole phone number? Can you explain how you remembered it? You may have tried to remember the number by visualising what you saw on the page – this is visual encoding. Alternatively, you might have tried to remember by repeating the number out loud to yourself (or silently, in your head) – this is acoustic. Short-term memory encoding is mostly acoustic, via rehearsal, as you may have just done with the phone number. Long-term memory is mostly semantic, meaning the information has some context or meaning. In this example, the phone number would not have had any context or meaning to attach to it.

The different ways in which information is encoded, or stored, has implications for the way in which it can be retrieved later. Miller (1956) famously claimed that adult short-term memory has a capacity limit of between five and nine items. This is frequently misinterpreted as five to nine individual items of information. What he actually meant was between five and nine information slots. Each of these information slots could vary greatly in the amount of information they contain, or could be as little as one digit or word, or it could be a complex image or event. It was this idea that led to the concept of chunking – that we can retain around seven items, or groups of items, in short-term memory for around thirty seconds. We often practise chunking of information in everyday life without realising. For example, do you ever have to stop and think about your phone number if someone reads it back to you using a different set of chunks to your usual way? For example, 077 8645 3389 versus 07 786 453 389.

6.6 Retrieval

Following encoding and storage, the third and final stage of memory processing is retrieval. Retrieval is the term used for accessing the information we have previously stored, because we need it in a current situation or for a current decision. Essentially, this is getting information out of storage. A bit like getting your Christmas decorations out of the loft.

The way in which information is retrieved differs between short-term and long-term memory systems. Short-term memories are usually stored and retrieved sequentially, meaning information is recalled in the order it is retained. For instance, if you memorise a list of words, you will likely recall it in the same order. Long-term memories are a bit more complex and are usually stored and retrieved according to their association or meaning – so not in a specific order. This is why sometimes you might only remember something when you see or experience a trigger that reminds you of it, like returning to the living room to remember what you went upstairs to get. This is a very good reason to study for exams in different places, so you don't develop strong associations between specific memories and your study space and can only recall the information when there.

6.7 Theories of Memory Structure

6.7.1 Multi-store Model

Atkinson and Shiffrin (1968) proposed the multi-store model of memory (Figure 6.1). You might also come across this referred to as the stages model or the modal model. The multi-store model states that memory is a flow of information through a three-stage memory system – a simple linear system. Atkinson and Shiffrin (1968) stated that information flows from sensory memory to short-term memory through attention via rehearsal.

Information flows from short-term to long-term memory, and back into short-term memory through encoding and retrieval. It is also important to note that the model includes maintenance of information in short-term memory, which would be achieved through continued rehearsal.

Sensory memories, according to Atkinson and Shiffrin's model, are transferred to short-term memory by attention. Short-term memory contains information that we are currently aware of. It is estimated it can be kept active for around thirty seconds. It is important to note the limited capacity limit of short-term memory. Memories in short-term memory are pushed out as new information enters. However, if information currently being held in short-term memory is attended to, it will be transferred to the next stage and stored in long-term memory.

Peterson and Peterson (1959) demonstrated short-term memory with a simple experiment. Participants were presented with three letters and a three-digit number combination. During a retention interval they were

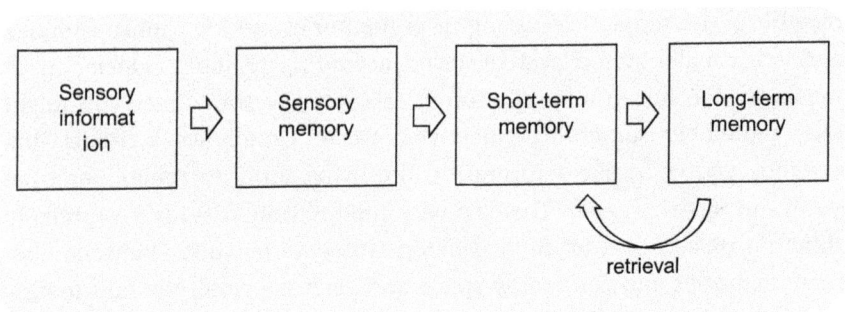

Figure 6.1 Atkinson and Shiffrin's (1968) multi-store model of memory.

asked to count backwards to prevent rehearsal. At a signal, participants were asked to recall the combination. The results showed that participants could not recall the combination, likely because they had been unable to attend to the information during the retention period. Think about the last time someone gave you a number or a code, did you repeat it to yourself until you found a pen or somewhere to type it? That is rehearsal, and a vital element of sensory memory.

The final stage of the multi-store model is long-term memory. Long-term memory is for the continued storage of information. Remember, this can last multiple decades. Long-term memory is thought to be largely away from awareness – meaning you have many memories retained or stored, but you are not constantly aware of them all. Long-term memories can be called into short-term or sensory memory whenever we need to access them, for example if we need them to inform a decision. In contrast to other memory stores that have a finite capacity, long-term memory does not have a known limit.

6.7.2 Working Memory

Baddeley (1986) viewed short-term memory as a mental scratchpad or temporary workspace, hence the name working memory. He proposed four components of working memory, which are the phonological loop, visuospatial sketchpad, episodic buffer and executive control system. Each component has its own specific purpose in the model (see diagram of the working memory model in Figure 2.1).

1. The phonological loop is responsible for speech-based information. This means any information that is verbal, written or anything else that involves communication through language.
2. The visuospatial sketchpad handles visual and spatial information. So this means anything you see or is in your spatial environment.
3. The executive control system is an attentional system that supervises and integrates information from the other systems.
4. The episodic buffer was added to a later version of the model. This is a third slave system that is thought to provide temporary integration of information before it is retrieved, reflected and manipulated by the central executive. Together, the episodic buffer and central executive integrate information from

the other systems. For instance, information entering your memory from watching and listening to a lecture enters the phonological loop in terms of what the lecturer is saying, and the visuospatial sketchpad in terms of what you can see on the slides. This is then integrated into one memory by the central executive.

It is generally agreed that memories can be stored differently according to how they are encoded. Paivio (1991) devised the dual coding hypothesis based on this knowledge. If you consider a concrete word – that means a word referring to something you can see or touch, like 'book' – the word book can be stored twice in long-term memory, once as a word and once as an image. However, for abstract words – those are words for things that you cannot touch or see, like 'thinking' – can only be stored as a word, because there is no visual image. Based on this idea, Paivio claimed that concrete words are remembered better than abstract words, because they are dual coded.

6.7.3 Levels of Processing Model

In 1972, Craik and Lockhart devised a model of memory that somewhat contradicts the earlier multi-store model. This model is termed the levels of processing model. The key to how this model organises memory is really in the name. The main concept of the levels of processing model is that the depth at which information is processed has a direct influence on how it is recalled. It is thought that certain types of encoding require more in-depth cognitive processes, and information that is encoded in these ways is more easily recalled. For instance, visual encoding of information is considered to be shallow. If we use the example of revising for an exam, visual encoding would be just looking at the lecture slides. Acoustic information is considered a deeper level of processing. This might include listening to the recorded lecture, as well as looking at the slides. Semantic processing is the deepest type of encoding. Semantic processing means to attach meaning to the information. So when you are revising the lecture content, this means you would take the time to think about what the information means and interpret it, rather than simply memorising the words being said. It is the semantic information you are going to recall more easily when it comes to the exam, because it has undergone deeper processing.

6.7.4 Anatomy of Memory

There are several key brain regions and structures that are very important in memory; however, there is no one location that we can call the memory area of the brain. Importantly, the areas of the brain that are activated by memory encoding and recall will differ depending on the type of memory. Many brain areas and structures are interconnected by complex networks to enable memory, but there are three key components for memory function. They are the hippocampus and the amygdala, as well as the cortex. Other regions including the cerebellum and basal ganglia are important in motor memories; the prefrontal cortex is very important in working memory, as well as consolidation and integration of memory information in general. This is important when it comes to decision-making.

The hippocampus is very important for memory. It is known to play an important role in memory formation, especially in the encoding of episodic and autobiographical memories. Evidence has shown the hippocampus to be involved in using memory to enable us to recognise new events and new experiences. The hippocampus is linked to the amygdala and is active in the encoding of emotional memory information. The amygdala is important for emotion. It is important to note that although we often refer to the hippocampus as a singular, we actually have two, one in each hemisphere. This is very beneficial in the case of damage to one of the hippocampi, because we are usually able to compensate with the other and see little impact on memory function. Damage to both however can result in problems forming new memories as well as problems accessing existing memories.

A lot of our early understanding of the role of the hippocampus in memory came from a case study, of patient HM (Scoville & Milner, 1957). HM received neurosurgery in an attempt to treat epilepsy. This surgery involved the removal of the medial structures of the temporal lobe, in both hemispheres. This included the hippocampus. HM's seizures stopped after the surgery, so in that sense it was a success. However, following the surgery he experienced severe memory impairment. HM was able to remember his friends and family, who he had known before the surgery well. He could remember information from his past. He did however struggle remembering information from the two years before the surgery. This is minor retrograde amnesia. Minor because he had no trouble remembering things beyond those two years, retrograde because we are talking about memories formed before the damage. His severe memory problems were in forming new memories. After the surgery he was never able to remember new people he met or recall events that had happened after the surgery. This is anterograde amnesia, anterograde because we are referring to information from after the damage.

The case of HM is an important one for memory science. The observations made in this one case were able to establish the fundamental principles of how memory functions are organised in the brain. Before HM it was assumed memory was widely distributed across the cortex, but the evidence from this case shows this is not the case.

6.7.5 Working Memory and Decision-Making

Working memory is the brain's vital ability to temporarily store information for use in complex processes, including decisions. Without the capacity for working memory, the integration of information required for decision-making would not be possible. Research evidence shows that better working memory leads to more efficient decision-making and working memory load can directly impact decision-making performance (Cui et al., 2015). Working memory is involved in a wide range of cognitive processes and is an important part of human brain function, but it is of particular relevance to how we make decisions.

Key Points

- Memory can be classified by fractations of duration: sensory, short-term and long-term, and by information type: semantic, episodic and autobiographical.
- Both encoding and retrieval are important for memory function.
- Without working memory, decision-making is not possible.

REFERENCES

Atkinson, R. C., & Shiffrin, R. M. (1968). Human memory: A proposed system and its control processes. In K. W. Spence & J. T. Spence (Eds.), *The psychology of learning and motivation* (Vol. 2, pp. 89–195). Academic Press.

Baddeley, A. (1986). *Working memory.* Clarendon Press/Oxford University Press.

Craik, F. I. M., & Lockhart, R. S. (1972). Levels of processing: A framework for memory research. *Journal of Verbal Learning and Verbal Behavior, 11*, 671–684. https://doi.org/10.1016/S0022-5371(72)80001-X

Cui, J. F., Wang, Y., Shi, H. S., Liu, L. L., Chen, X. J., & Chen, Y. H. (2015). Effects of working memory load on uncertain decision-making: Evidence from the Iowa Gambling Task. *Frontiers in Psychology, 6*, 162. https://doi.org/10.3389/fpsyg.2015.00162

Miller, G. A. (1956). The magical number seven, plus or minus two: Some limits on our capacity for processing information. *Psychological Review, 63*(2), 81–97. https://doi.org/10.1037/h0043158

Paivio, A. (1991). Dual coding theory: Retrospect and current status. *Canadian Journal of Psychology / Revue canadienne de psychologie, 45*(3), 255–287. https://doi.org/10.1037/h0084295

Peterson, L. R., & Peterson, M. J. (1959). Short-term retention of individual verbal items. *Journal of Experimental Psychology, 58*(3), 193–198. https://doi.org/10.1037/h0049234

Scoville, W. B., & Milner, B. (1957). Loss of recent memory after bilateral hippocampal lesions. *Journal of Neurology, Neurosurgery, and Psychiatry, 20*(1), 11–21. https://doi.org/10.1136/jnnp.20.1.11

7

· · · · · · ·

Decision Networks

7.1 The Underlying Anatomical Networks of Decision-Making

Through earlier chapters, it has been established that in contrast to other cortical regions, including visual and motor cortices, the prefrontal cortex (PFC) remains relatively poorly understood in terms of the fine details of its structure. Research evidence has pointed towards ordered arrangements of connections in a topographic form, similar to what has long been accepted as a common feature of brain connectivity (e.g. Goldman-Rakic, 1988; Kondo & Witter, 2014; Sesack et al., 1989). Fairly recent evidence has confirmed the existence of reciprocal connections between the PFC and other cortical regions, showing consistent alignment of connections (Triplett et al., 2009). However, these conclusions have been somewhat questioned by recent evidence highlighting further complexities when explored on a smaller scale (Agster & Burwell, 2009; Bedwell & Tinsley, 2018). To date, there is no universally accepted anatomical structural map of the PFC.

7.2 PFC: Temporal Cortex Network

In the first systematic study of its kind, Bedwell et al. (2014, 2015, 2017; Bedwell & Tinsley, 2018) set out to identify the anatomical arrangement of connections from the PFC in rats. The highly systematic approach, which had not previously been applied in the field, enabled the creation of a baseline and confident understanding of anatomical connectivity in this pathway, which led to question assumptions of reciprocity and aligned point-to-point mapping which had been accepted until this point.

The co-application of retrograde and anterograde tract tracers to precise predetermined sites in the PFC allowed for the visualisation of both input and output connections. For accuracy, the study implemented a statistical analysis to determine whether PFC connections that were visualised displayed an ordered arrangement comparable to present understanding of anatomical organisation elsewhere in the cortex. The three-dimensional location of each retrogradely and anterogradely labelled cell was measured.

The statistical analysis of these precisely measured locations was a novel method, that allowed accurate conclusions to be made in regard to the relationship between input and output connections from the same PFC source. The results of this study sparked interest because they clearly demonstrated that input and output connections from the PFC to temporal cortices follow different orders of connectivity – they appeared to move in opposing directions when moving from medial to lateral in the PFC (Figure 7.1).

Importantly, the findings from this rigorous study did identify a general, broad-scale, ordered pattern of connections in line with previous conclusions. The observed opposing order of connections for anterograde and

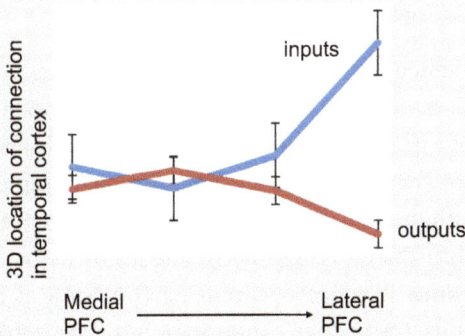

Figure 7.1 The location of PFC connections in temporal cortex.

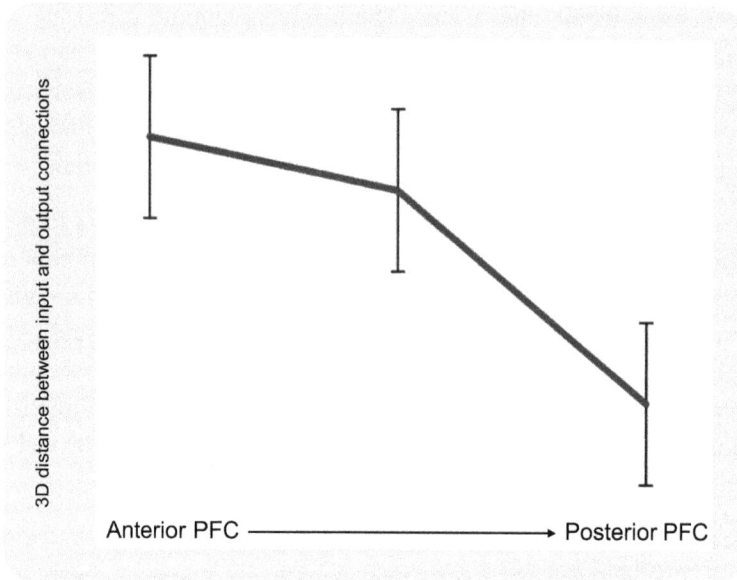

Figure 7.2 Comparison of input and output connections from the PFC.

retrograde connections was seen consistently across axes of orientation and was replicated.

A wider area of the PFC was studied, from the most anterior pole to the most posterior aspect of the PFC. Calculated Euclidean distances between labelled inputs and outputs were statistically compared and a significant difference found between them from anterior to posterior. Notably, a significant difference was only found in anterior PFC regions, indicating that anatomical connections become more aligned with one another the further posterior one moves in the PFC (Figure 7.2).

It was also observed that divergence increases as connections move from anterior to posterior in the PFC. It was concluded that PFC connections to temporal cortex follow a gradient from anterior to posterior. This observation is consistent with functional suggestions of an organisational gradient (Taren et al., 2011) and aligns with the idea that more anterior regions of the PFC are more involved in the more abstract cognitive processes we associate with high-order executive functions like decision-making (Christoff et al., 2009). These anatomical findings formed an important basis on which further research has begun to explore the network requirements for more abstract processing, perhaps being less alignment and reciprocity than previously thought.

7.3 The Unique Organisation of PFC Connections

The evidence on a fairly broad scale points towards a general ordered pattern of connections consistent with what would be expected based on other cortical regions. However, it is when investigations move to a much finer resolution, including multiple sources in the same cytoarchitectural regions, that we begin to see the unique complexities of PFC organisation (Bedwell & Tinsley, 2018).

Only through finer-scale analysis of connectivity, that is overlooked on a larger scale, can we appreciate the details of PFC organisation. Findings established that the input and output connections on a fine scale are unaligned, consistent with what can be seen at a broader scale. Despite an overall similar appearance of connectivity, when viewed on a finer scale, unique structural properties were identified. Where input and output connections appeared to move in opposing directions on a broader scale, when viewed in more detail it can be seen that this relationship is more complex than this. The way in which the relationship between inputs and outputs change differs or is not consistent between the two. That is, inputs change at a greater rate than outputs.

On a broad scale, there is a general theme of reciprocity between inputs and outputs – in that labelling from the two can be found in the same general cytoarchitectural region. However, when visualised in greater detail it appears they are less reciprocal than assumed.

The connectivity matrix in Figure 7.3 shows the spatial relationship between input and output connections from prefrontal to temporal cortices. The accompanying graph visualises how this relationship changes across the PFC.

The graph shows us that as the source in the PFC moves from medial to lateral, the median distance between input and output connections decreases. Interestingly, although we know that the prelimbic cortex, the

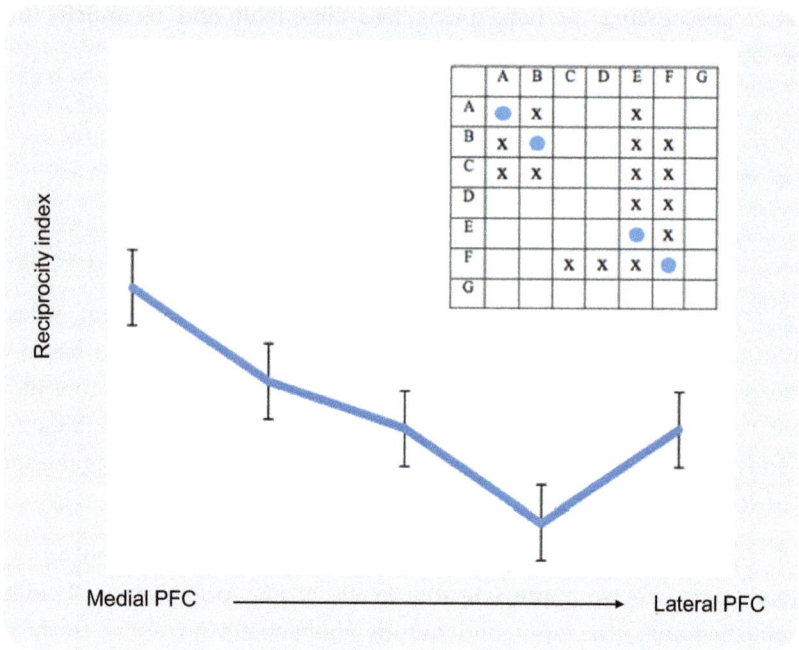

Figure 7.3 Reciprocity of PFC connections.

most medial region in the rat, is the most reciprocal in terms of connectivity, it also has the most widespread connectivity in this network.

DISCUSSION POINT

Consider the possible functional advantages of differential organisation of connections in the PFC.

7.4 Other PFC Networks

This chapter has focused so far on one of many networks involving the PFC. Evidence from other pathways shows similar structural properties. For instance, connections between the PFC and sensory-motor cortices show similar evidence for broad-scale ordering, where input and output connections appear to move in opposing directions (Bedwell et al., 2015), but this ordering is revealed to be more complex when visualised on a finer scale (Bedwell & Tinsley, 2018).

7.5 The Underlying Functional Networks of Decision-Making

Unravelling the complexities of anatomical brain networks is key to gaining a full understanding of complex functions, including decision-making. Alongside developments in knowledge surrounding anatomical structural properties in PFC pathways, in recent years research into the structure of functional networks associated with executive functions and decision-making has grown rapidly.

The relatively new field of computational neuroscience allows for the combination of methodologies from biological and psychological sciences, and mathematical modelling to enable the identification of network properties that otherwise cannot be seen. Computational methods in functional neuroscience have the unique ability to reveal the dynamic nature of complex networks. Whilst attempting to establish a detailed network model down to the single neuron level is not currently possible, and experts claim this to be an unrealistic goal (Sporns, 2013), investigating at a macroscale is possible and can provide valuable information to improve current understanding of human decision-making.

7.6 Determining Functional Connectivity

In recent years, the application of graph theory to data derived from EEG studies has begun to reveal previously undescribed properties of functional networks related to cognitive processes, including various aspects of decision-making. To determine structural properties in this way, connectivity matrices are constructed whereby two nodes (in this case these are usually the locations of EEG electrodes on the scalp) are considered connected in the functional network if they produce a significantly correlated amplitude at a given point in time (corresponding to an event-related potential). Such matrices may be constructed in a weighted format, whereby the weight of an edge (connection between two nodes) corresponds to the size of the correlation coefficient. A greater weight corresponds to a more relevant connection. Or, connectivity matrices may be represented in a binary format, whereby 1 = a correlation coefficient above a specified threshold and 0 = a correlation coefficient below a threshold.

A recent study (Katwa et al., in preparation) applied this strategy to EEG data derived from Dufau et al. (2017), representing functional activity during a lexical decision-making task. Correlation matrices were constructed to represent the connectivity pattern at 280 ms post stimulus. Through the application of graph measures, it was established that there was a greater degree in more frontal areas as a whole in comparison to posterior regions. This is as expected, with prior knowledge of prefrontal importance in decision-making. More notably, the observed degree appears to follow a gradient with the frontal regions, indicating an increase in degree as nodes move from anterior to posterior frontal regions. This is the case for both weighted and binary networks.

These findings suggest that the frontal pole is less highly connected, therefore perhaps less heavily involved, in lexical decision-making tasks. These observations are consistent with anatomical evidence for an anterior to posterior gradient of connectivity within the PFC identified by Bedwell et al. (2017).

7.7 The Future of Functional Network Analysis of Decision-Making

The above is just one example of how combining knowledge from different fields can aid in the development of knowledge. Many more studies have been and continue to be carried out, furthering our understanding of how complex functions are structured. With this knowledge, future developments will enable the identification of changes to these structural properties through developmental stages, between populations, in neurodegeneration and in disease. As established in earlier chapters, decision-making deficits are associated with a range of disorders, including schizophrenia, depression and autism. Although we have this knowledge, we do not yet have the detailed understanding of decision-making network structure that is required to fully understand these differences. When such differences are fully understood, and only then, can we hope to develop successful interventions, treatments and maybe even prevention.

Key Points

- Cortical connections are generally considered to be reciprocally organised.
- Evidence indicates the PFC connectivity, while similar to other cortical regions on a broad scale, is organised differently on a finer scale.
- Both anatomical and functional studies reveal evidence for a gradient of connectivity in prefrontal networks.

REFERENCES

Agster, K. L., & Burwell, R. D. (2009). Cortical efferents of the perirhinal, postrhinal, and entorhinal cortices of the rat. *Hippocampus, 19*(12), 1159–1186. https://doi .org/10.1002/hipo.20578

Bedwell, S. A., Billett, E. E., Crofts, J. J., MacDonald, D., & Tinsley, C. J. (2015). The topology of connections between rat prefrontal and temporal cortices. *Frontiers in Systems Neuroscience, 9*(80). https://doi.org/10.3389/fnsys.2015.00080

Bedwell, S. A., Billett, E. E., Crofts, J. J., & Tinsley, C. J. (2014). The topology of connections between rat prefrontal, motor and sensory cortices. *Frontiers in Systems Neuroscience, 8*(177). https://doi.org/10.3389/fnsys.2014.00177

Bedwell, S. A., Billett, E. E., Crofts, J. J., & Tinsley, C. J. (2017). Differences in anatomical connections across distinct areas in the rodent prefrontal cortex. *European Journal of Neuroscience, 45*(6), 859–873. https://doi.org/10.1111/ejn .13521

Bedwell, S. A., & Tinsley, C. J. (2018). Mapping of fine scale rat prefrontal cortex connections: Evidence for detailed ordering of inputs and outputs connecting the temporal cortex and sensory-motor regions. *European Journal of Neuroscience, 48*(3), 1944–1963. https://doi.org/10.1111/ejn.14068

Christoff, K., Keramatian, K., Gordon, A. M., Smith, R., & Madler, B. (2009). Prefrontal organization of cognitive control according to levels of abstraction. *Brain Research, 1286*, 94–105. https://doi.org/10.1016/j.brainres.2009.05.096

Dufau, S., Grainger, J., Midgley, K., & Holcomb, P. (2017). The kilo-word ERP database (lexical decision). Retrieved from https://osf.io/72b89

Goldman-Rakic, P. S. (1988). Topography of cognition: Parallel distributed networks in primate association cortex. *Annual Review of Neuroscience, 11*, 137–156. https://doi.org/10.1146/annurev.ne.11.030188.001033

Katwa, G., Bedwell, S. A., Rogers, J., & Brookes, M. (in preparation). The functional network structure of lexical decision-making.

Kondo, H., & Witter, M. P. (2014). Topographic organization of orbitofrontal projections to the parahippocampal region in rats. *The Journal of Comparative Neurology, 522*(4), 772–793. https://doi.org/10.1002/cne.23442

Sesack, S. R., Deutch, A. Y., Roth, R. H., & Bunney, B. S. (1989). Topographical organization of the efferent projections of the medial prefrontal cortex in the rat: An anterograde tract-tracing study with Phaseolus vulgaris leucoagglutinin. *Journal of Comparative Neurology, 290*, 213–242. https://doi.org/10.1002/cne.902900205

Sporns, O. (2013). The human connectome: Origins and challenges. *Neuroimage, 80*, 53–61. https://doi.org/10.1016/j.neuroimage.2013.03.023

Taren, A. A., Venkatraman, V., & Huettel, S. A. (2011). A parallel functional topography between medial and lateral prefrontal cortex: Evidence and implications for cognitive control. *Journal of Neuroscience: The Official Journal of the Society for Neuroscience, 31*(13), 5026–5031. https://doi.org/10.1523/JNEUROSCI.5762-10.2011

Triplett, J. W., Owens, M. T., Yamada, J., Lemke, G., Cang, J., Stryker, M. P., & Feldheim, D. A. (2009). Retinal input instructs alignment of visual topographic maps. *Cell, 139*(1), 175–185. https://doi.org/10.1016/j.cell.2009.08.028

8

• • • • • • •

Decision-Making Styles and Models of Decision-Making

8.1 Categorising Decisions

Decision and decision-making are broad terms, that encompass processes from the simple decision to move a limb, to a complex life-changing decision to move to a new country, and everything in between. In order to understand processes that vary so greatly but have common underpinnings, it is necessary to be able to break them down into categories, types or styles of decision.

8.2 Decision-Making Styles

Decision-making style can be defined as a habitual pattern – an automatic or routine behaviour that has been learnt, often through reward. Individuals use this learnt habitual behaviour, their own decision-making style, in the decisions they make. We can consider decision-making style an individual characteristic model of perceiving and responding to decision-making tasks in varied contexts.

Early researchers in the concept of decision-making style suggested that decision-making style could be defined by the amount of information gathered, for example from past experiences, and the number of alternatives considered when a person is making a decision. In the 1970s and 1980s, several researchers focused their aims on the way in which individuals gather and process information to make a decision. For example, McKenney and Keen (1974) suggested that people bring their established habits and strategic modes of thinking to organising information in a certain way, affecting how they process it. They suggested that, when making a decision, there are two main styles in which people can be categorised.

1. We will structure the problem in terms of a method likely to lead to a solution.
2. We will use a strategy of trial and error.

Later on, Driver (1979) proposed that decision-making style is something that is learned. The key differences in styles in which we make decisions involve the amount of information considered during a decision, and the number of alternatives identified when reaching a decision. Based on this conclusion, Driver suggested each individual might have a primary and secondary decision-making style, rather than the previously implied one. Which style is used at a given time might depend on the circumstances, the importance placed on the decision or the amount of prior knowledge we have to use in informing the decision and its consequences.

Building on the understanding of Driver (1979), Harren (1979) proposed there are not two, but three decision-making styles.

1. Dependent: dependent decision makers are described as projecting responsibility onto others. They are likely to see themselves as not being in control.
2. Rational: rational decision makers assume personal responsibility and take a more deliberate and logical approach to decisions.
3. Intuitive: intuitive decision makers also take personal responsibility for their decisions and rely more on what we might describe as an internal hunch or gut feeling.

8.3 The General Decision-Making Style Questionnaire

The development of theories of decision-making style in the latter half of the twentieth century led to the development of a means of measuring decision-making style. Some of the first researchers to create a quantifiable measure of decision-making style were Scott and Bruce (1995). They created the General Decision-Making Style questionnaire (GDMS). This measure was designed to assess how individuals approach decision situations, and distinguishes between five decision-making styles. The aim is that each individual who completes the measure can be categorised into one prominent style.

1. A rational decision-making style emphasises a thorough search for a logical evaluation of alternatives.
2. An avoidant decision-making style emphasises postponing and avoiding decisions.
3. A dependent decision-making style emphasises a search for advice and direction from others.
4. An intuitive decision-making style emphasises a reliance on hunches and feelings.
5. A spontaneous decision-making style emphasises a sense of immediacy, and a desire to get through the decision-making process as soon as possible.

The questionnaire consists of twenty items, all of which are answered on a three-point Likert scale ranging from 1 (false) to 3 (true). Each item is a statement, to which an individual should respond on the scale to indicate how true the statement is for them. The twenty items can be broken down into the five subscales. Higher scores in relation to any of the five subscales are indicative of a tendency towards that decision-making style. An individual may have tendencies towards more than one decision-making style. The authors (Scott & Bruce, 1995) suggested most people have some inclination towards all five styles. The GDMS questionnaire is considered a valid measure of decision-making style.

8.4 Your Decision-Making Style

Why not try the GDMS out for yourself? Below are the twenty items of the GDMS. For each statement, you should rate on the scale how much it applies to you. Put a rating of 1 if it is false, a rating of 2 if it is sometimes true and a rating of 3 if it is true. Keep a note of your ratings for each statement.

1. I will make sure that I have all the facts before I make a decision.

   1 = false 2 = sometimes true 3 = true

2. When I make a decision, I do what feels right.

   1 = false 2 = sometimes true 3 = true

3. I often ask other people to help me make important decisions.

   1 = false 2 = sometimes true 3 = true

4. I don't like making decisions, so I try to avoid it.

   1 = false 2 = sometimes true 3 = true

5. I make decisions quickly.

   1 = false 2 = sometimes true 3 = true

6. I make decisions in a slow, logical way.

   1 = false 2 = sometimes true 3 = true

7. When I make a decision, I rely on my instincts.

   1 = false 2 = sometimes true 3 = true

8. I don't make big decisions without talking to other people first.

   1 = false 2 = sometimes true 3 = true

9. I usually won't make an important decision until I'm forced to do so.

   1 = false 2 = sometimes true 3 = true

10. I don't think too much about the decisions that I make.

    1 = false 2 = sometimes true 3 = true

11. Making decisions requires careful thought.

    1 = false 2 = sometimes true 3 = true

12. A decision doesn't need to make sense, it just needs to feel right.

    1 = false 2 = sometimes true 3 = true

13. When I need to make an important decision, I like to have someone point me in the right direction.

    1 = false 2 = sometimes true 3 = true

14. I try to put off making important decisions because thinking about them makes me feel uneasy.

    1 = false 2 = sometimes true 3 = true

15. When I need to make an important decision, I just do what seems natural at the moment.

    1 = false 2 = sometimes true 3 = true

16. I consider all of my options before making a decision.

    1 = false 2 = sometimes true 3 = true

17. I rely on my inner feelings when making decisions.

    1 = false 2 = sometimes true 3 = true

18. When I make a decision, I rely on other people's advice.

1 = false 2 = sometimes true 3 = true

19. I usually make important decisions at the last minute.

1 = false 2 = sometimes true 3 = true

20. I often make impulsive decisions.

1 = false 2 = sometimes true 3 = true

These statements can be categorised into the five decision-making styles. Add your score for each decision-making style to get a score for each.

The total scores you have for each list indicate how much you align with that decision-making style, based on your answers to the GMDS questionnaire.

If A is your highest number, you are a *systematic* decision-maker.
If B is your highest number, you are *intuitive*.
If C is your highest number, you are *dependent*.
If D is your highest number, you are *avoidant*.
If E is your highest number, you are *spontaneous*.

If you have equally high scores in more than one area, this indicates that you likely regularly use more than one decision-making style.

8.5 Systematic Decision-Making Style

If you are a systematic decision-maker, you make decisions slowly and carefully. You make sure that you have all the facts before you proceed. You carefully consider all of your options in order to make the best choice.

8.6 Intuitive Decision-Making Style

If you are an intuitive decision maker, when you need to make a decision, you tend to rely on your feelings. You make decisions based on 'hunches' and instinct. You do what feels right at the moment.

8.7 Dependent Decision-Making Style

If you are a dependent decision maker, you find it hard to make decisions without other people's support. When you need to make an important decision, you ask other people for advice and direction. You might be uncomfortable making decisions alone.

8.8 Avoidant Decision-Making Style

If you are an avoidant decision maker, you don't like making important decisions. They make you feel uneasy and uncomfortable, so you try to avoid making decisions as much as possible. You only make important decisions when you're pressured to do so.

8.9 Spontaneous Decision-Making Style

If you are a spontaneous decision maker, you make important decisions quickly and easily. You are often impulsive. You don't spend much time thinking or worrying about your decisions. You just make a choice and see what happens.

Whatever your results are, it is important to remember that one decision-making style is not better than another, and there is no right or wrong way to make decisions. The whole concept is actually quite subjective. Each decision-making style described by the GDMS has its strengths and weaknesses, and each might work well in different situations or circumstances. Skilled decision makers are able to use more than one style. In real life, most of us do. They're able to be flexible and use the style that best fits the situation.

DISCUSSION POINT

Consider the limitations of the GDMS and categorisation of people into decision-making styles.

In line with previous suggestions that a person can move between decision-making styles, or have a primary and secondary style, research evidence also shows relationships between Scott and Bruce's (1995) decision-making styles. A 2015 study carried out by Bavolar and Orosova investigated relationships between decision making styles (Bavolar & Orosova, 2015). The study revealed a higher reported use of the rational decision-making style correlated positively with the dependent decision-making style. A possible explanation for this relationship could be that an effort to seek advice from other people is part of a rational process – so the two styles become intertwined. The study also found a negative correlation between the rational and the avoidant style. This could mean that the people who tend to avoid decisions are also those who are spontaneous in decision-making. A possible explanation could be that in situations where it is necessary to decide, they want to do it as soon as possible. These relationships found between decision-making styles highlight that individuals have a dominant style even if they tend to use more than one decision-making style. However, it should be noted these findings are based on correlational analyses and cannot determine a cause and effect. Interestingly, the same study (Bavolar & Orosova, 2015) also identified a negative correlation between avoiding decision-making and poor mental health. That is, the more a person fits into the avoidant decision-making style, the poorer their mental health. Again, this is a correlation, so we cannot assume causation, but it may be sensible to assume that those who avoid making decisions may experience higher levels of stress, for example, as a result. Or such individuals may avoid the decision to seek help for poor mental health, thus causing further deterioration. Further research is needed to fully understand these relationships.

8.10 Individual Differences in Decision-Making

We have established through earlier chapters that decision-making is far from being a simple cognitive process. Decision-making is one of the most complex and multifaceted elements of human cognition that we have knowledge of, and it varies in nature from person to person. Each individual has their own decision-making style, which may change over time and depending on the situation. Each individual also has their own unique life experiences, learning and knowledge that they bring to each decision to be made.

As humans, we have vast individual differences. It is what makes us interesting and not copies of one another. Individuals are all different in their approaches to decisions, their goals and how these play a role in the consideration of outcomes. These differences are in part attributed to biological factors which differ between individuals, and also environmental factors – which are never identical for two people; even identical twins raised in the same family will have some variations in their environmental experiences.

8.11 Individual Differences in Approaches to Decision-Making

Research has established already that individuals have different decision-making styles, a general way of making decisions that applies to many decision processes. Individuals are also thought to take varied approaches to the way they make decisions based on other influential factors including their general cognitive style, specific personality traits and significant life experiences (both positive and negative).

Personality traits are thought to have a significant impact on decision-making style. For instance, highly conscientious people tend to make more deliberate and carefully considered decisions, and extroverts have been found to be generally more impulsive in their decision-making (McCrae & Costa, 1999). On the other hand, less risky decision-making is associated with higher levels of neuroticism (DeYoung et al., 2007).

A range of environmental factors are thought to influence decision-making and decision-making style. Evidence shows a possible effect of family, culture, education and exposure to varied life experiences. Notably, it is thought that cultural differences in values and cultural norms could play a significant role in individual differences in decision-making (Oyserman et al., 2002). European Americans show more individualistic traits, whereas Chinese Americans are more collectivist. These distinct cultural differences are linked to an effect on relationality and cognitive style, which impacts decision-making.

Differences in the biological underpinnings of decision-making, including the action of neurotransmitters and the development of complex brain networks associated with decision-making, are also thought to play an important role in the individual differences of approaches to decision-making (see chapters 5, 7, 9 and 11).

The individual differences we see in the way we approach and make decisions are a product of many contributory factors including biological variation, environmental influence, personality traits and past experiences. It is partly due to this wide range of influencing factors that understanding the complexities of human decision-making is a vast task.

Key Points

- The GDMS categorises people into five decision-making styles: rational, avoidant, dependant, intuitive, spontaneous.
- People may move between decision-making styles, or have a primary and secondary style.
- A range of factors are thought to impact decision-making style, including family, culture, education and various life experiences.

REFERENCES

Bavolar, J., & Orosova, O. (2015). Decision-making styles and their associations with decision-making competencies and mental health. *Judgment and Decision Making, 10*(1), 115–122.

DeYoung, C. G., Quilty, L. C., & Peterson, J. B. (2007). Between facets and domains: 10 aspects of the Big Five. *Journal of Personality and Social Psychology, 93*(5), 880–896. https://doi.org/10.1037/0022-3514.93.5.880

Driver, M. J. (1979). Individual decision making and creativity. In S. Kerr (Ed.), *Organizational behavior* (pp. 55–91). Grid Publishing.

Harren, V. A. (1979). A model of career decision making for college students. *Journal of Vocational Behavior, 14*(2), 119–133. https://doi.org/10.1016/0001-8791(79)90065-4

McCrae, R. R., & Costa, P. T. (1999). A five factor theory of personality. In L. A. Pervin & O. P. John (Eds.), *Handbook of personality: Theory and research* (2nd ed., pp. 139–153). Guilford.

McKenney, J. L., & Keen, P. G. W. (1974). How managers' minds work. *Harvard Business Review, 53*(3), 79–90.

Oyserman, D., Coon, H. M., & Kemmelmeier, M. (2002). Rethinking individualism and collectivism: Evaluation of theoretical assumptions and meta-analyses. *Psychological Bulletin, 128*(1), 3–72. https://doi.org/10.1037/0033-2909.128.1.3

Scott, S. G., & Bruce, R. A. (1995). Decision-making style: The development and assessment of a new measure. *Educational and Psychological Measurement, 55*(5), 818–831. https://doi.org/10.1177/0013164495055005017

9

• • • • • • •

Development

Like every part of the human body, the brain follows a clear and reliable timeline for development across the lifespan, beginning as a fetus and taking us all the way through to old age. Various stages of development are key to specific processes and cognitive functions, many of which can be reliably pinpointed to a specific age in development. These are often described as critical periods of development. Some stages of brain development are key in the development of executive functions and decision-making. Unlike some more basic functions that may have only one critical period of development, executive functions, due to their complexity, undergo several key stages of development throughout the lifespan.

The main stages of human brain development can be broken down into five specific periods: prenatal, newborn, childhood, adolescence and adulthood.

9.1 The Prenatal Stage

9.1.1 Neuron Creation, Cell Migration, Myelination

The formation of neurons starts with the embryo. By five weeks after conception, cells begin dividing rapidly to form the approximately 100 billion neurons that a baby will have at birth. Once they are formed, the neurons in the developing brain start to migrate towards specific locations, and synapses (connections) begin to form between them. This is the beginnings of the 100 trillion synapses that will be present in the adult brain.

Interestingly, evidence shows that some external factors like pressure, temperature and the physical movements of the fetus in the uterus can stimulate synapse formation (Ackerman, 1992). The process of axon myelination also begins in utero, thought to be at around sixteen weeks' gestation (Gaillard et al., n.d.). This is an important part of brain development when considering the development of complex processes like decision-making, as such cognitively demanding tasks depend greatly on efficiency, for which myelinated axons are vital.

9.1.2 What Can Influence Brain Development in Utero?

We often think of environmental influences on development as things that happen to a child after they are born, but there are many ways in which environment can influence brain development whilst a fetus is developing:

- maternal nutrition;
- maternal stress;
- maternal infection; and
- maternal drug exposure.

All of the above are known to be associated with the onset of neurodevelopmental disorders. Barker and Osmond (1986) defined the fetal programming hypothesis (Barker hypothesis) in which they stated future health conditions are determined by early environmental influences during prenatal and postnatal development. As many neurodevelopmental disorders are associated with abnormalities in the development of cortical networks, it is plausible to conclude that such environmental influences could have a direct impact on the early development of synapses.

9.2 The Newborn Stage

9.2.1 Basic Survival and Innate Reflexes and Improvement of Information Transfer

At the time of birth, a baby's brain can weigh up to 400 grams and contains around 100 billion neurons, similar to that of an adult brain. It is important to acknowledge that most of what we think of as brain and cognitive development is to do with the development and strengthening of connections between cells that already exist, not the growth of new neurons. At this point, many synapses have developed in areas involved in basic survival and innate reflexes making these areas well-developed. These are the connections that allow a baby to survive through basic bodily functions, but also to carry out simple behaviours, like crying when they require a need to be met like feeding or changing.

During the first year, white matter volume increases up to around 16 per cent (Knickmeyer et al., 2008). Neurons associated with sensory functions like vision and hearing quickly become myelinated during the newborn stage and therefore improve their information transfer soon after birth. This can be observed in a baby's rapid development in visual processing and responding to familiar auditory cues, such as their caregiver's voice. If you observe a new-born baby in the first few days and weeks of life, you will see their ability to recognise the sound of their mother's voice improve very quickly. There is a clear survival and evolutionary advantage to prioritising development in this ability, as it allows the baby to start to associate specific people with safety and security.

During the early weeks and months of life, the human brain is rapidly changing. The brain of a baby is constantly producing new synapses, essentially overproducing connections. Many of these new connections formed in the early days will not be used later on. Connections which are not needed will later be removed through a process known as synaptic pruning.

DISCUSSION POINT

Consider different factors, both environmental and biological, that could influence neurodevelopment in infants.

9.3 The Childhood Stage

9.3.1 Critical Stages of Development, Synaptogenesis and Synaptic Pruning

Historically, Piaget (1952) described several stages to cognitive development during childhood. Piaget's four stages are: the sensorimotor stage up to around two years old; the preoperational stage from around two to seven years old; the concrete operational stage from around seven to eleven years old; and the final stage, the formal operational stage from eleven years old into adulthood. Piaget stated that decision-making skills develop and begin to emerge through all four of these stages. Piaget's thoughts on the stages of cognitive development somewhat align with what we see in the stages of development from a more biological perspective, with the contemporary knowledge and understanding we have of nervous system development.

In the early years of childhood, humans are greatly limited in decision-making capability, not only by biological underpinnings but also in limited general cognitive abilities – which are by nature linked back to biological development. During the period of development Piaget described as preoperational, approximately between the ages of two and seven years, decisions children make usually largely focus on their own personal needs and desires. Often, decisions during this time are thought to be centred on immediate gratification and concrete knowledge. Children at this age struggle to comprehend the abstract ideas and thinking that would be required to carry out more complex decisions involving forward planning or problem-solving. This lack of ability in complex decision-making coincides with our understanding of biological development in brain regions and networks associated with such decisions. However, there is some evidence to suggest young children around the age of five can begin to make decisions that consider future consequences (Mischel et al., 1989).

Later in childhood, during the time Piaget referred to as the concrete operational stage, approximately between the ages of seven and eleven years, children's decision-making abilities begin to become more sophisticated, showing greater capabilities in more complex and abstract decisions. This phenomenon is particularly relevant if we look at improvement in risky decision-making tasks (Blair et al., 2001). During these years, most children develop the ability to use logic in their decision-making. Cognitive development elsewhere enables the application of problem-solving to complex decisions, as well as the consideration of multiple factors, for example in considering multiple consequences, benefits and costs based on experience and knowledge. During this stage children also start to develop an understanding of external factors such as social norms, which begin to guide decisions – moving on from the earlier stages where most decisions were based on personal needs and desires. This progress in the complexity of decision-making aligns with the progress seen in the underlying biological development.

Throughout childhood, the human brain is rapidly changing. The brain reaches about 90 per cent of its adult size by the time a child is seven years old. During the early years, the brain goes through three important processes.

1. Synaptogenesis is the formation of synapses or connections between neurons, allowing information to be transmitted between them. This process is vital for forming complex brain networks and ultimately the development of cognitive functions like decision-making.
2. Synaptic pruning is a process that begins in childhood and continues through to adulthood. During the process we call synaptic pruning, existing connections or synapses that are developed are removed when they're not needed. This process of removing unnecessary or unused connections improves efficiency and connectivity in the brain.
3. Myelination is a very important developmental process. The process of the formation of the myelin sheath begins with the fetus and continues throughout life. Most major changes in myelination occur within the first year of life. Myelination improves the efficiency and speed of information transfer.

Research evidence demonstrates that complex cognitive skills, required for complex decision-making, for example cognitive shifting, develop rapidly in early childhood (Kirkham et al., 2003; Zelazo et al., 1996). Where three-year-old children will struggle with a task-switching activity, most five-year-olds can successfully perform it. This development coincides with the development of working memory during this stage; where four-year-old children struggle with working memory tasks such as the self-ordered searching task (Luciana & Nelson, 1998), seven- and eight-year-old children succeed.

9.3.2 Role of Social and Moral Development

It is important we recognise that decision-making development is not happening in isolation. Also during childhood, we develop both socially and morally. This is especially important during the childhood years. Kohlberg (1981) suggested a staged model of moral development. They proposed that in the early stages of development children follow rules, and they do so to avoid punishment or to gain rewards – not because of their morality. For instance, a young child may decide not to steal a toy from another child because they have learnt they will be punished if they do so. The young child is not making this decision because of a belief it is morally wrong to steal. An older child, however, may decide not to steal a toy from another because of their moral judgement. The older child is making the decision because they believe it is morally wrong to steal; there is less emphasis on the fear of punishment in their decision-making process. According to Kohlberg's (1981) model, as the child matures, so does their moral and ethical reasoning, and this begins to play a greater role in decisions.

9.3.3 Parental Roles

Many external factors have the potential to influence decision-making development. Parental modelling is one factor considered to be highly influential, primarily due to the nature in which young children learn many behaviours – through observation (Xiao et al., 2011). Parents and how they behave in terms of making decisions play an important role in decision-making development during childhood. As well as through modelling, parents set boundaries and rules, implement punishments, rewards and consequences and in many cases help and guide children through decision-making processes. It is through parental influence and learning processes implemented by parents or caregivers, such as positive reinforcement, that some synaptic connections are strengthened.

9.3.4 Education

Throughout childhood, we are influenced beyond our parents or caregivers. Most children experience school or other educational environments, which also play a substantial role in cognitive development. Decision-making development is particularly relevant here; classroom discussions, learning processes, problem-solving and interacting with other children provide ample opportunity to practise and refine decision-making skills. Children will also be guided, in a similar way to that of parental influence, in their decision-making by their teachers.

9.3.5 Peers

Friends and classmates become increasingly important to brain development as children get older, in some ways replacing the earlier parental influence. Interactions with peers during a period described as middle childhood can start to shape the decisions children make. This may be a result of peer pressure but also through a desire to meet social norms – to fit in. Often, as children begin to get older and reach pre-teen and adolescent years, they begin to prioritise the values of their peer group in their decisions over the values they have been taught by their parents or caregivers (Silva et al., 2016).

9.3.6 Understanding Consequences

In the early years of childhood, it is often not useful to teach a behaviour using consequences, as very young children lack the capacity to understand (Arnall, 2019). Before the age of six, children struggle to imagine the consequences of planned actions. For instance, a young toddler is unlikely to understand that if they decide not to pick up their toys, they won't get chocolate. An older child is more likely to comprehend that there are consequences to decisions, so deciding not to put their toys away when asked will result in no chocolate. Developing the ability to understand consequences to behaviours is an essential element in the development of complex decision-making. Positive reinforcement strengthens certain behaviours, whilst negative consequences encourage the adjustment of behaviours, or learning of new ones.

The same concept applies to teaching a dog a new behaviour. They lack the brain development to understand consequences – think of your pet dog like a human toddler. Teaching a new puppy new behaviours using positive reinforcement is a great strategy but introduce the idea of punishment or negative consequences and everything is likely to go wrong. Your cute puppy cannot understand that if he decides to chew the new rug he will get yelled at, so use this strategy and he will keep doing it. He can quickly learn though that if he sits nicely when you ask, he gets a delicious treat. So soon he will be more likely to sit nicely and less likely to decide to chew the rug because it gets no positive reward.

Figure 9.1

9.4 The Adolescence Stage

9.4.1 Risky Decision-Making, Prolonged Development of Prefrontal Cortex and Executive Functions, Synaptic Pruning, Hormone Changes

It is easy to view teenagers as essentially mini-adults. But as far as brain development goes, especially executive functioning, this could not be further from the truth. Adolescent brains are a work in progress. During the teenage years, there are numerous hormone changes within the body, and the brain is no exception to this. In the years between childhood and adulthood, dopamine levels increase, especially in the prefrontal cortex (PFC). Dopamine is an important neurotransmitter when it comes to decision-making. Dopamine activity in the PFC specifically is linked to many cognitive processes, including working memory, stress responses, emotional control and executive functions (see Chapter 5). An increase in availability of dopamine increases the excitability of GABAergic interneurons in the PFC.

In addition to hormonal changes, the myelination and synaptic pruning that was happening during childhood continues during this time. This is very prominent in the PFC. The continued myelination, along with continued pruning of connections that are not needed in the PFC, improves the efficiency of information processing and connectivity between this region and others. Increased connectivity and efficiency of networks within and including the PFC during this time is strengthening the processes associated with prefrontal function, like decision-making, forward planning and goal direction.

The physical size of the human brain does not continue to grow as teenagers get older, having reached its largest physical size by around the age of 11–14. However, under the surface, growth is still ongoing.

9.4.2 Myelination and Synaptic Pruning

In early childhood, the brain went through a period of large-scale synaptic pruning. It has long been understood that the brain of a baby essentially overproduces connections and removes those which are not needed or used. It is only relatively recently that research evidence revealed the same process happening again in adolescence (Crews et al., 2007). This process in adolescence, is particularly focussed on the frontal cortex, particularly the PFC. By the time we have exited adolescence into adulthood, it is estimated that our brains hold approximately 41 per cent fewer synaptic connections than we had as an infant (Abitz et al., 2007).

Adolescence is a crucial period of development for many physical, social and cognitive changes. Changes in terms of decision-making development are considered critical during this time. The development of decision-making during the teenage years involves the structural changes of the brain, as well as social development and the impact of hormonal and emotional changes. In terms of structural development, the physical changes we see in the PFC during adolescence have a direct impact on decision-making behaviours. It is important to note that the complexity of the PFC and its prolonged development mean that structural development is not uniform. That is, development in some areas outpaces others. For instance, evidence shows that the limbic system develops more quickly during adolescence than the PFC, resulting in an imbalance when it comes to decision-making (Casey et al., 2008). The more matured limbic system in comparison to the PFC can lead to heightened emotional reactivity, making adolescents more likely to make impulsive decisions (Casey et al., 2008).

Cognitive control, as established in Chapter 2, is essential for executive functions, including decision-making. The cognitive processes that allow for cognitive control (attentional flexibility, inhibitory control and working memory) enable us to evaluate different outcomes, consider consequences and to make choices that allow us to work towards goals and plans. This ability is continuing to improve through adolescence, and as it does so, teenagers begin to be able to make more strategic decisions, perhaps focused more on forward plans (Best & Miller, 2010) and becoming less impulsive.

When considering cognitive development, it is important we recognise that as humans we do not develop in a vacuum. Whilst genetically predetermined biological development is taking place, our cognitive development, including that of executive functioning and decision-making, is also influenced by external factors. During adolescence, an influential external factor is that of peer influence.

9.4.3 Social Influence and Impulsivity

As children become adolescents, they start to become less influenced by parents/caregivers and more influenced by peers. This level of influence will later decrease into adulthood, as social influence becomes less important. Peer influence can have a significant impact on decision-making during this stage, and thus also on the ongoing decision-making development. The value teens place on peer approval can sometimes override rational decision-making during this time, resulting in making decisions in an attempt to achieve social approval, rather than one's own reasoning (Albert et al., 2013).

During the adolescent stage, cognitive control begins to mature; however, executive function is still undergoing rapid change and teens are prone to impulsive behaviours, which can link to the way in which they make decisions. Reward systems in the brain that started to develop in childhood become highly active during adolescence (Eshel et al., 2007), resulting in often reward-seeking behaviours, even where they might be fully aware of potential negative consequences. This leads to the risky behaviours often linked to teenagers.

The strong influence of social approval during adolescence, as well as tendencies for impulsivity, are linked to the increased risk-taking behaviour often observed in this age group. It is not necessarily the case that adolescents are not able to make as rational decisions as their older counterparts, but more that social approval holds a higher value at this stage of development.

9.4.4 Moral Reasoning

Considering consequences for oneself and others is a vital element of complex decision-making, and a skill that adolescents develop into adulthood. During this stage we begin to develop the ability to critically evaluate situations and consider consequences of different decisions before we make them. Kohlberg's (1981) stages of moral development model states that adolescents will progress from avoiding punishment in their decisions during childhood to making decisions based on their own moral judgement in adulthood. In other words, we progress from not hurting someone because we do not want to get punished, to not hurting someone because we think it is morally wrong to do so.

9.5 The Adulthood Stage

9.5.1 Prefrontal Cortex and Associated Functions Are Considered Fully Developed around the Age of Twenty-Five

The most important brain region to become what we consider fully developed during adulthood is the PFC. The PFC is heavily involved in high-order complex functions like decision-making, forward planning, goal direction, inhibition and other executive functions. It is when we reach around the age of twenty-five that we often consider the PFC to be fully developed, and our ability in decision-making to have reached its full potential (Arain et al., 2013). It is at this stage that the other influences of adolescence that may have been causing greater risky decision-making will subside, and our own moral judgement in decisions will usually take precedent. The full maturity of our decision-making abilities is not however the end of changes to our decision-making processes. Through adulthood, new experiences and knowledge continue to contribute to the decisions we make.

It is important to note that although we consider the human brain to be fully matured by around the age of twenty-five, our brains do not stop changing through our whole lifetime. We are always forming new connections and continue to weaken others. The brain is never static and will continue to develop in some way every day of our lives.

9.6 The Ageing Brain

9.6.1 Continued Strengthening and Development of New Connections through Adulthood

Throughout the adulthood years, connections will continue to be strengthened where utilised and weakened where they are not. New experiences and learned information will continue to influence decisions as they are made, with the use of working memory.

9.6.2 Cognitive Decline

As the brain begins to enter old age, structural and functional changes are known to occur, which have varying impacts on cognitive function. This is known as cognitive decline. Cognitive decline in later life is a natural part of the ageing process and encompasses many cognitive processes, including executive functions like decision-making. Understanding how cognitive decline plays a role in decision-making for healthy older adults is becoming increasingly important as our population ages.

Decision-making is one of the most complex cognitive processes of the human brain, involving the integration of information from multiple systems. Therefore, it makes sense that decision-making is particularly susceptible to cognitive decline. Memory is a faculty which is significantly impacted by the gradual process of cognitive decline, and has a direct impact on decision-making abilities. Working memory in particular, vital for complex decision-making, is known to show deficits associated with cognitive decline (Park & Reuter-Lorenz, 2009). As working memory abilities begin to decrease with age, the ability to consider all relevant information or relevant past experiences in a decision begins to diminish. Ultimately, this leads to individuals making choices they would not have in previous years, before cognitive decline had this effect. The extent to which cognitive decline occurs and impacts decision-making is dependent on individual differences, varying within and between populations (Randhawa & Varghese, 2023).

9.6.3 Age-Related Changes in Risk Perception

We established that adolescents are more likely to make risky decisions due to social influence, and this settles down by adulthood. Older adults have been shown to show opposing behaviours to teenagers in regard to risk taking, in that they tend to avoid potential risky decisions more so than those in earlier adulthood (Mather, 2012). This may be a self-preservation mechanism – as we become older and more vulnerable, we begin to avoid risky decisions more, in order to prevent poor choices (Löckenhoff & Carstensen, 2007).

Key Points

- Human brain development can be broken down into five stages: prenatal, newborn, childhood, adolescence and adulthood.
- Decision-making development occurs in some form at all five stages of development.
- The prefrontal cortex and its associated functions are considered fully mature around the age of twenty-five.

REFERENCES

Abitz, A., Nielsen, R. D., Jones, E. G., Laursen, H., Graem, N., & Pakkenberg, N. (2007). Excess of neurons in the human newborn mediodorsal thalamus compared with that of the adult. *Cerebral Cortex, 17*(11), 2573–2578. https://doi.org/10.1093/cercor/bhl163

Ackerman, S. (1992). *Discovering the brain.* National Academies Press (US).

Albert, D., Chein, J., & Steinberg, L. (2013). Peer influences on adolescent decision making. *Current Directions in Psychological Science, 22*(2), 114–120. https://doi.org/10.1177/0963721412471347

Arain, M., Haque, M., Johal, L., Mathur, P., Nel, W., Rais, A., Sandhu, R., & Sharma, S. (2013). Maturation of the adolescent brain. *Neuropsychiatric Disease and Treatment, 9*, 449–461. https://doi.org/10.2147/NDT.S39776

Arnall, J. (2019, 18 February). *When do children understand consequences?* https://judyarnall.com/2019/02/18/when-do-children-understand-consequences

Barker, D., & Osmond, C. (1986). Infant mortality, childhood nutrition, and ischaemic heart disease in England and Wales. *Lancet, 327*(8489), 1077–1081. https://doi.org/10.1016/s0140-6736(86)91340-1

Best, J. R., & Miller, P. H. (2010). A developmental perspective on executive function. *Child Development, 81*(6), 1641–1660. https://doi.org/10.1111/j.1467-8624.2010.01499.x

Blair, R., Colledge, E., & Mitchell, D. (2001). Somatic markers and response reversal: Is there orbitofrontal cortex dysfunction in boys with psychopathic tendencies? *Journal of Abnormal Child Psychology, 29*(6), 499–511. https://doi.org/10.1023/a:1012277125119

Casey, B. J., Jones, R. M., & Hare, T. A. (2008). The adolescent brain. *Annals of the New York Academy of Sciences, 1124*(1), 111–126. https://doi.org/10.1196/annals.1440.010

Crews, F., He, J., & Hodge, C. (2007). Adolescent cortical development: A critical period of vulnerability for addiction. *Pharmacology Biochemistry and Behavior, 86*(2), 189–199. https://doi.org/10.1016/j.pbb.2006.12.001

Eshel, N., Nelson, E. E., Blair, R. J., Pine, D. S., & Ernst, M. (2007). Neural substrates of choice selection in adults and adolescents: Development of the ventrolateral prefrontal and anterior cingulate cortices. *Neuropsychologia, 45*(6), 1270–1279. https://doi.org/10.1016/j.neuropsychologia.2006.10.004

Gaillard, F., Zhang, B., Ibrahim, D., et al. (n.d.). Normal myelination [reference article]. Radiopaedia.org. https://doi.org/10.53347/rID-5776

Kirkham, N. Z., Cruess, L., & Diamond, A. (2003). Helping children apply their knowledge to their behaviour on a dimension-switching task. *Developmental Science, 6*(5), 449–467. https://doi.org/10.1111/1467-7687.00300

Knickmeyer, R. C., Gouttard, S., Kang, C., Evans, D., Wilber, K., Smith, J. K., Hamer, R. M., Lin, W., Gerig, G., & Gilmore, J. H. (2008). A structural MRI study of human brain development from birth to 2 years. *Journal of Neuroscience, 28*(47), 12176–12182. https://doi.org/10.1523/JNEUROSCI.3479-08.2008

Kohlberg, L. (1981). *Essays on moral development: The philosophy of moral development.* Harper & Row.

Löckenhoff, C. E., & Carstensen, L. L. (2007). Aging, emotion, and health-related decision strategies: Motivational manipulations can reduce age differences. *Psychology and Aging, 22*(1), 134–146. https://doi.org/10.1037/0882-7974.22.1.134

Luciana, M., & Nelson, C. A. (1998). The functional emergence of prefrontally-guided working memory systems in four- to eight-year-old children. *Neuropsychologia, 36*(3), 273–293. https://doi.org/10.1016/s0028-3932(97)00109-7

Mather, M. (2012). The emotion paradox in the aging brain. *Annals of the New York Academy of Sciences, 1251*(1), 33–49. https://doi.org/10.1111/j.1749-6632.2012.06471.x

Mischel, W., Shoda, Y., & Rodriguez, M. (1989). Delay of gratification in children. *Science, 244*(4907), 933–938. https://doi.org/10.1126/science.2658056

Park, D. C., & Reuter-Lorenz, P. (2009). The adaptive brain: Aging and neurocognitive scaffolding. *Annual Review of Psychology, 60*, 173–196. https://doi.org/10.1146/annurev.psych.59.103006.093656

Piaget, J. (1952). *The origins of intelligence in children.* International Universities Press.

Randhawa, S. S., & Varghese, D. (2023). *Geriatric evaluation and treatment of age-related cognitive decline.* StatPearls. www.ncbi.nlm.nih.gov/books/NBK580536

Silva, K., Chein, J., & Steinberg, L. (2016). Adolescents in peer groups make more prudent decisions when a slightly older adult is present. *Psychological Science, 27*(3), 322–330. https://doi.org/10.1177/0956797615620379

Xiao, L., Bechara, A., Palmer, P. H., Trinidad, D. R., Wei, Y., Jia, Y., & Johnson, C. A. (2011). Parent-child engagement in decision making and the development of adolescent affective decision capacity and binge drinking. *Personality and Individual Differences, 51*(3), 285–292. https://doi.org/10.1016/j.paid.2010.04.023

Zelazo, P. D., Frye, D., & Rapus, T. (1996). An age-related dissociation between knowing rules and using them. *Cognitive Development, 11*(1), 37–63. https://doi.org/10.1016/S0885-2014(96)90027-1

10

• • • • • • •

The Role of Childhood Experiences in Decision-Making

10.1 Influences on Cognitive Development

Childhood is known to be an important time period for cognitive development. It is widely accepted that there are critical periods for various facets of development in cognitive processing, with more complex abilities such as those described by executive function going through several critical periods. Many aspects of cognition and the way such processes develop in the growing brain are genetically determined. That is, our ability to form a memory, to retrieve a memory, to integrate information from multiple brain regions, is all determined by how our brain is structured anatomically and how it functions on a basic level physiologically. This is basically the same for everyone. However, a growing body of evidence demonstrates a significant role of environment and experience in the development of complex cognitive abilities.

Decision-making is arguably one of the most important cognitive functions; after all, we make countless decisions every day in order to function as a human being. Despite its importance, exactly how decision-making develops, what influences that development, and especially the role of experiences and childhood environment is relatively poorly understood. There is a growing focus within the fields of psychology and neuroscience to fill this gap (Bechara, 2004; Bedwell et al., 2023; Clark, 2010; Clark et al., 2008).

10.2 The HPA Axis

The hypothalamic–pituitary–adrenal (HPA) axis is a major neuroendocrine system with great importance when it comes to understanding executive function. It is composed of a series of complex interactions between three main components or structures: the hypothalamus, pituitary gland and adrenal gland. It is the complex physiological interactions between these three composing structures that we refer to as the HPA axis. This complex system is implicated in the body's reaction to stress, the regulation of digestion, mood and emotions, immune response and has also been linked to sexuality.

In terms of explaining decision-making development, the relationship between the HPA axis and our reaction to stress is most relevant. The HPA axis is also implicated in mood and emotion, which is also important in considering the underpinnings of decision-making. The stress response is the main role of the HPA axis. How this physiological response is activated during development can have lasting effects far beyond that which we may directly relate to stress.

Anatomical connections from key structures such as the amygdala and hippocampus to the prefrontal cortex are understood to contribute to the activation of the HPA axis. Sensory information arriving at the amygdala is processed and projected onwards to several regions and structures involved in the response to fear and stress. In the hypothalamus, fear signal impulses activate the HPA axis. When activated through this process, the neurotransmitter cortisol is produced. So, cortisol is a product of HPA axis activation through fear or stress. As a follow-on effect from this, increased cortisol production during the fear or stress response results in an increased availability of glucose. This is the body's way of preparing us for a fight or flight response to a stressful situation – the additional glucose facilitates whichever physical action we take.

Cortisol is a vital neurotransmitter. It acts with adrenaline to enable the formation of memories associated with emotional events. These are referred to as flashbulb memories. It is thought that the main purpose of a flashbulb memory is to enable us to make informed decisions, based on past experience, to avoid similar threatening or stress-inducing situations in the future. The role of the HPA axis and cortisol are key in decision-making, especially when we consider in more depth how this system influences development.

10.3 Role of the HPA Axis in the Stress Response and Decision-Making

It is clear and widely understood that the HPA axis is important in the regulation of the stress response, thus enabling appropriate responses to environmental stresses or threatening situations. The HPA axis is developed in early childhood and is dependent largely on exposure to different stressors. Experiencing abnormal levels of stress in childhood has been shown to contribute significantly towards the physiological dys-regulation of the HPA axis. Specifically, interruption or changes to the typical development of the HPS axis can result in chronically elevated levels of cortisol and reduced synaptic plasticity, which has a somewhat knock-on effect when it comes to the development of high-order functions, like decision-making.

Synaptic plasticity refers to the constant changes that occur in terms of neural connectivity in response to experiences, learning and memory formation. Part of synaptic plasticity is the strengthening and development of new connections. This dynamic nature of brain connectivity has a direct effect on the way complex networks in the brain develop and how those networks then go on to interpret environmental cues, new experiences and stress responses, from childhood through to adulthood. In other words, synaptic plasticity and factors that impact the formation of networks have an effect on decision-making in the future. In a scenario where a child's ability to learn how to regulate their response to stress is compromised, their capacity to regulate their own response in the future becomes compromised, often resulting in permanent changes from the norm. A clear example of this effect can be seen in repeated exposure to violence during childhood. Such experiences result in repeated activation of the cortical networks involved in the stress response.

Our knowledge from neuroanatomy and physiology tells us that neural connections that are repeatedly activated become strengthened over time, which can eventually result in a functional change. In the case of repeated activation of the stress response from exposure to violence in childhood, a permanent functional change may be the child's (and subsequent adult) response to stress and fear. Current knowledge in this area stems from cases of severe childhood trauma; however, it may be the case that other prolonged or repeated childhood experiences, such as sibling aggression or aggression with peers, can impact the development of the HPA axis and thus the manner in which children learn to deal with stressors and make decisions in stressful situations into adulthood (Bedwell et al., 2023).

Further evidence suggests that symptom complexity increases linearly with the amount of exposure to adverse or traumatic experiences, much like a dose–response relationship. If similar relationships exist with other childhood experiences, we may expect functions associated with the HPA axis to be more greatly impacted in those who experience higher levels of sibling aggression, peer aggression and other childhood stressors. The functional outcome of childhood stressors is thought to be largely dependent on developmental timing, likely due to the constant structural changes occurring in the developing cortex. Complex decision-making is particularly sensitive to the effects of synaptic plasticity and the prolonged development of prefrontal cortical connectivity, meaning it is affected by the ability of connections to vary in strength of connections over time.

10.4 Trauma and Adverse Childhood Experiences

10.4.1 What Is an Adverse Childhood Experience?

Defined by Young Minds (2018) as 'highly stressful, and potentially traumatic, events or situations that occur during childhood and/or adolescence', an adverse childhood experience (ACE) may be an isolated event or an ongoing experience. A traumatic experience can be defined as an ACE, but an ACE is not necessarily considered to be traumatic. Some examples of ACEs include loss of a parent, exposure to domestic violence, emotional, physical or sexual abuse, living with someone who abuses drugs or alcohol, living with someone who has gone to prison and living with someone with a mental illness. A 2014 study (Bellis et al., 2014) estimated that almost half (47 per cent) of all adults experienced at least one ACE.

10.4.2 The Impact of ACEs on Executive Function

Adverse childhood experiences, including trauma, are known to impact the development of executive functions, as a result of lasting physiological and anatomical effects, specifically on cortical and subcortical structures including the hippocampus, amygdala and medial prefrontal cortex (Anda et al., 2006; McCrory et al., 2012; Pechtel & Pizzagali, 2011; Wilson et al., 2011). These structures are highly associated with the regulation of mood. The medial prefrontal cortex is of particular relevance as it is associated with high-order cognitive functions including decision-making, problem-solving, forward planning and social inhibition (Alvarez & Emory, 2006; Fuster, 2001; Kolb, 1984). The same region is also known to play a role in the manifestation of psychological disorders involving decision-making deficits such as schizophrenia. Further, a high prevalence of specific cognitive abnormalities involving decision-making deficits have been reported in adolescents who experienced childhood trauma in some form (Dauvermann & Donohoe, 2018).

The impact of trauma on cognitive development is widely accepted, although understanding of ACEs in a wider sense remains to be fully established. Exposure to ACEs during childhood is understood to be related to a range of cognitive functions and underlying physiology (Dannlowski et al., 2012; Hart & Rubia, 2012). Evidence consistently shows a lasting effect of ACEs on the biology of the brain. Specially, repeated traumatic experiences are linked to cortical and subcortical abnormalities associated with anxiety and mood disorders (Anda et al., 2006; Pechtel & Pizzagali, 2011).

Importantly, the impact early childhood trauma has on executive function development is thought to depend on a number of factors. This can include type of trauma, how prolonged the exposure to trauma was and the age at which the traumatic experience occurred, or age of onset if prolonged.

We know there are critical periods of development within childhood, both physically and mentally. During critical developmental times, when neural pathways are being formed, functional and structural networks are significantly altered by traumatic events. Evidence shows the outcome of traumatic experiences in terms of cognitive impact to be largely dependent on the timing of the traumatic event (Cook et al., 2005). Complex decision-making undergoes a critical period of development, which is dependent on synaptic plasticity, making it particularly vulnerable to external influence (Crews et al., 2007). During development, pathways can become overdeveloped or under-developed, dependent on the environmental influence at play.

10.4.3 Effects of Childhood Trauma on Inhibition

It is widely understood across psychology and neuroscience that childhood trauma is linked to impairments in cognition, and specifically in executive function and working memory; processes thought to underlie psychological disorders including depression and schizophrenia. Recent investigations (Bedwell & Hickman, 2023) have explored the possibility of a relationship between childhood trauma, psychopathic traits and response inhibition, a vital element of the decision-making process. Findings indicate that child-hood trauma does not predict susceptibility to psychopathology traits or a deficit in response inhibition, contrary to what may be expected based on understanding of the negative impact of trauma. These findings form an important basis on which to build a further understanding of the complex consequences of childhood trauma exposure. Specifically, in terms of understanding how specific cognitive functions may be influenced or not, and by providing a clearer understanding of how psychopathic traits develop and how decision-making may be involved. The current evidence base shows that trauma has significant effects on cognitive functioning, particularly executive functions. However, where effects have been observed these are often mediated by diagnosis of a psychiatric disorder.

It is important to note, that although research evidence points towards a relationship between childhood trauma, or ACEs, and decision-making deficit, that relationship is far from being a linear one. It is not a simple case of every child who experiences a traumatic event will go on to develop a deficit in decision-making, or a particular style of decision-making. If this were the case, we would have a much clearer understanding of the process. As with all human functions and processes, within the brain and elsewhere, nothing happens in isolation. There are many other factors involved.

Although there is evidence for the role of trauma in executive function, the specific role, along with the influence of response inhibition, remains to be fully understood. Research evidence implies that trauma-exposed children have poorer performance on executive function tasks, that may play a role in later behaviours, for example decisions in response to aggression.

10.5 Role of Childhood Experiences in Decision-Making

A growing body of literature shows that exposure to stressful events during childhood has a significant impact on a range of high-order cognitive functions and their underlying physiological organisation, and that these effects can be seen into adulthood. As established earlier in this chapter, it is understood that traumatic experiences in childhood may have a lasting neurobiological effect. Additionally, research evidence consistently shows that repeated stress and trauma result in neurobiological abnormalities in cortical and subcortical regions, specifically regions associated with the mediation of anxiety and mood. Relevant structures known to be implicated include the hippocampus, amygdala and the prefrontal cortex.

It is well established that the prefrontal cortex is involved in a number of high-order, complex cognitive processes, including decision-making, problem-solving, forward planning and social inhibition among others. The same region is also implicated in a range of neurological and psychological deficits including psychosis, depression and autism. With this knowledge, it is reasonable to conclude that physiological changes impacting the development of prefrontal cortex connectivity would have an effect on these functions and may even play a role in the development of such disorders and divergencies.

Given the research that has shown repeated stress and childhood trauma to have a long-lasting impact on cortical and subcortical regions of the brain – specifically those associated with anxiety and mood, like the hippocampus, and as we mentioned, the prefrontal cortex – it could be possible that experiences not often thought of as particularly traumatic, like experiencing childhood sibling aggression, also have a similar effect, thus influencing resultant behaviours.

It is known that the human prefrontal cortex undergoes a period of prolonged development and is not fully developed until early adulthood, thought to be around the age of twenty-five. It has been established that the organisation of prefrontal cortex connections is highly complex and differs in terms of specific structural components in comparison to other more understood cortical regions. Such complexity and structure provide increased opportunity for multiple factors to influence development. It is known that there are critical times throughout cognitive development during which the formation of neural pathways can be changed by traumatic events. It remains unclear, however, how common childhood experiences, such as sibling aggression, could contribute to the most complex human cognitive processes associated with the prefrontal cortex, such as decision-making.

10.6 Stress in Childhood

Scott and Bruce (1995) described decision-making style as a learned response in a specific decision-making context. It's understood that a variety of decision-making styles exist with the general population and that there are many environmental and physiological factors that could contribute to the decision-making style an individual displays. There is an established effect of early childhood stresses caused by parental attachment or neglect on cognitive development, and a known link between childhood experiences, for example, attachment and decision-making style. Similarly, decision-making style has been found to be associated with well-being and mental health. But there is limited evidence to support the role of trauma in decision-making style. However, the impact of childhood stresses occurring as a result of other negative childhood experiences, such as the sibling aggression we've been talking about, remains relatively little understood. Specifically, the available literature has not examined in depth how childhood experiences, stress or aggression affects decision-making in adulthood. This is what my colleagues and I are currently investigating.

DISCUSSION POINT

How might positive experiences in childhood play a role in decision-making in adulthood?

10.7 Current Research Developments

Projects currently being conducted between King's College London, Keele University and Manchester Metropolitan University are working on improving understanding of the relationship between early experience and executive functions like decision-making. One recent study (Bedwell et al., 2023) identified a significant association between experiences of childhood sibling aggression and decision-making styles reported in adulthood. Specifically, the retrospective study identified a link between using sibling aggression to maintain dominance during childhood and an avoidant decision-making style reported in adulthood. This means that people who, as children, used aggression to gain control of resources or possessions, were more likely to be avoidant in their decision-making style as an adult. This finding is of particular interest because one might expect the opposite. Understanding of avoidance as a personality trait is that it is commonly associated with inhibition and introversion, fear of criticism and social anxiety. Based on this, it would be logically to expect victims of sibling aggression to become avoidant adults, especially because avoidance is often linked to childhood emotional neglect and abusive childhood environments – it is a trait usually linked to being a victim, not a perpetrator. By exploring the motivations behind behaviours linked to avoidance, it may be the case that the link is not being a victim or perpetrator itself, but the reasoning behind perpetrator behaviour that leads to avoidant decision styles.

Being the victim of sibling aggression in childhood has been linked to increased levels of anxiety and depression in adulthood and an increased likelihood of engaging in aggressive behaviour in later intimate relationships (Bowes et al., 2014; Dantchev et al., 2018; Dantchev & Wolke, 2019; Perkins & Meyers, 2020). This latest research extends knowledge with regard to the psychological consequences of sibling aggression.

It is understood that there is a higher likelihood of neurobiological abnormalities in individuals of repeated childhood trauma. When this is considered alongside these findings, one may suggest that using aggression as a way to maintain dominance and control resources has an impact on prefrontal cortex development and, therefore, on exhibited high-order processes such as decision-making. It is clear that experiences of sibling aggression are linked to decision-making style. However, exploring how the underlying neuroanatomical and neurophysiological development network structures are influenced and contribute to these exhibited behaviours requires further investigation.

10.8 Risky Decisions

Findings from Bedwell et al. (2023) revealed a significant relationship between self-reported sibling aggression and decision-making style reported in adulthood, which suggests experiences of sibling aggression impact decision-making style. Therefore, one may expect to find a similar role of sibling aggression in real-time risky decision-making performance. However, when this was measured as part of the same study, findings showed this not to be the case – experiences of sibling aggression were not related to how people performed on a risky decision-making task (the Iowa gambling task). The findings indicate that sibling aggression does not predict risky decision-making, but does impact decision-making style. This suggests that the types of decisions made, or how they are made, might be influenced by sibling aggression during developmental stages in childhood, but the riskiness of decisions made are not. It's important to note here that this is just the beginning, and risky decisions are only one type of decision from many – more studies need to be conducted to be able to conclude more thoroughly how sibling aggression influences decision-making as a whole.

Interestingly, the findings from Bedwell et al. (2023) do not align with our existing knowledge of how trauma influences decision-making and executive function. This suggests that it is inaccurate to group experiences of sibling aggression in with other traumatic experiences, because their influence could be quite different.

10.9 Post-traumatic Growth

Historically, research has focused on the negative impacts of trauma, such as repression and amnesia of the traumatic event, as well as impacts on everyday memory (Guez et al., 2011; Hulbert & Anderson, 2018). Although understanding negative consequences of trauma is important, it is equally valuable to understand the possibility of positive outcomes. In recent years, there has been a movement towards establishing an understanding of the positive cognitive and neuropsychological changes that can result from traumatic events (Hussain & Bhushan, 2011).

Post-traumatic growth (PTG) refers to a positive change in perception of life and of self, following an adverse event (Tedeschi & Calhoun, 1996). There are said to be five main aspects of this positive change (Jin et al., 2014; Tedeschi & Calhoun, 2004):

1. Personal strengths
2. Enhanced relationships
3. Positive spiritual alterations in mindset
4. Optimism of possibilities
5. Gratefulness for life

Post-traumatic growth gives people increased confidence in themselves, as well as psychological maturity (Harmon & Venta, 2021). It is important to note that PTG does not lessen the trauma, or the emotional impact of the trauma (Hussain & Bhushan, 2011). Jin et al. (2014) describe people as gaining more wisdom and insight into their level of functioning with PTG. Although sometimes described as a positive outcome of trauma, it is important to recognise that trauma by nature has a negative impact on an individual, and the development of PTG does not make trauma a positive experience.

Constructive cognitive processing is thought to be an important factor in the development of PTG, in that the trauma may trigger a person to begin thinking constructively about themselves and the world regarding the event (Zhang et al., 2018). These cognitive processes are critical for PTG to develop as they enable people to think about the negative event in a useful way, by accepting and positively reinterpreting what has happened (Wong & Yeung, 2017). Intrusive rumination and brooding over a traumatic event are likely to lead to continued distress and a negative outlook on life, whereas deliberate rumination and reflection are linked to positive, happier outcomes (Stockton et al., 2011). This suggests that actively thinking, or deliberately pondering, about the past event may have adaptive cognitive qualities and can lead to PTG (Stockton et al., 2011). Zhang et al. (2018) found that that intentionally thinking about the trauma can lead to positive growth.

The development of PTG is facilitated by cognitive processes, including executive functions (Wong & Yeung, 2017). If a person has greater ability in executive functions, including decision-making, they may be more likely to be able to develop PTG after experiencing trauma. For example, Eren-Koçak & Kiliç (2014) found that PTG was related to higher executive functions, specifically personal growth. Kira et al. (2020) found that recovering from certain types of trauma could be aided by PTG, resulting in a reduction in working memory and inhibition deficits usually associated with trauma.

Key Points

- Environment and experience play an important role in cognitive development.
- Developmental influences on the HPA axis may have a significant contribution to decision-making development due to action on cortisol production.
- ACEs impact the development of executive functions, through physiological and anatomical effects.
- The development of PTG is facilitated by cognitive processes, including executive functions (Wong & Yeung, 2017).

REFERENCES

Alvarez, J. A., & Emory, E. (2006). Executive function and the frontal lobes: A meta-analytic review. *Neuropsychology Review, 16*(1), 17–42. https://doi.org/10.1007/s11065-006-9002-x

Anda, R. G., Feliti, V. J., Bremner, J. D., Walker, J. D., Whitfield, C., Perry, B. D., Dube, S. R., & Giles, W. H. (2006). The enduring effects of abuse and related adverse experiences in childhood. *European Archives of Psychiatry and Clinical Neuroscience, 256,* 174–186. https://doi.org/10.1007/s00406-005-0624-4

Bechara, A. (2004). The role of emotion in decision-making: Evidence from neurological patients with orbitofrontal damage. *Brain and Cognition, 55*(1), 30–40. https://doi.org/10.1016/j.bandc.2003.04.001

Bedwell, S. A., Harrison, N., Fradley, S., & Brooks, M. (2023). The role of sibling aggression during childhood in decision-making during adulthood. *Current Psychology, 43,* 2264–2276. https://doi.org/10.1007/s12144-023-04475-7

Bedwell, S. A., & Hickman, C. (2023). Effects of childhood trauma in psychopathy and response inhibition. *Development & Psychopathology, 35*(2), 724–729. https://doi.org/10.1017/S0954579421001863

Bellis, M. A., Hughes, K., Leckenby, N., Perkins, C., & Lowey, H. (2014). National household survey of adverse childhood experiences and their relationship with resilience to health-harming behaviors in England. *BMC Medicine, 12,* 72. https://doi.org/10.1186/1741-7015-12-72

Bowes, L., Wolke, D., Joinson, C., Lereya, S. T., & Lewis, G (2014). Sibling bullying and risk of depression, anxiety, and self-harm: A prospective cohort study. *Pediatrics, 134*(4), e1032–e1039. https://doi.org/10.1542/peds.2014-0832

Clark, L. (2010). Decision-making during gambling: An integration of cognitive and psychobiological approaches. *Philosophical Transactions of the Royal Society B: Biological Sciences, 365*(1538), 319–330. https://doi.org/10.1098/rstb.2009.0147

Clark, L., Bechara, A., Damasio, H., Aitken, M. R., Sahakian, B. J., & Robbins, T. W. (2008). Differential effects of insular and ventromedial prefrontal cortex lesions on risky decision-making. *Brain: A Journal of Neurology, 131*(Pt 5), 1311–1322. https://doi.org/10.1093/brain/awn066

Cook, A., Spinazzola, J., Ford, J., Lanktree, C., Blaustein, M., Cloitre, M., DeRosa, R., Hubbard, R., Kagan, R., Liautaud, J., Mallah, K., Olafson, E., & van der Kolk, B. (2005). Complex trauma in children and adolescents. *Psychiatric Annals, 35*(5), 390–398. https://doi.org/10.3928/00485713-20050501-05

Crews, F., He, J., & Hodge, C. (2007). Adolescent cortical development: A critical period of vulnerability for addiction. *Pharmacology Biochemistry and Behavior, 86*(2), 189–199. https://doi.org/10.106/j.pbb.2006.12.001

Dannlowski, U., Stuhrmann, A., Beutelmann, V., Zwanzger, P., Grotegerd, D., Domschke, K., Hohoff, C., Ohrmann, P., Bauer, J., Lindner, C., Postert, C., Konrad, C., Arolt, V., Heindel, W., Suslow, T., & Lenzen, T. (2012). Limbic scars: Long-term consequences of childhood maltreatment revealed by functional and structural MRI. *Biological Psychiatry, 71*(4), 286–293. https://doi.org/10.1016/j.biopsych.2011.10.021

Dantchev, S., & Wolke, D. (2019). Trouble in the nest: Antecedents of sibling bullying victimization and perpetration. *Developmental Psychology, 55*(5), 1059–1071. https://doi.org/10.1037/dev0000700

Dantchev, S., Zammit, S., & Wolke, D. (2018). Sibling bullying in middle childhood and psychotic disorder at 18 years: A prospective cohort study. *Psychological Medicine, 48*(14), 2321–2328. https://doi.org/10.1017/S00332917003814

Dauvermann, M. R., & Donohoe, G. (2018). The role of childhood trauma in cognitive performance in schizophrenia and bipolar disorder: A systematic review. *Schizophrenia Research: Cognition, 16,* 1–11. https://doi.org/10.1016/j.scog.2018.11.001

Eren-Koçak, E., & Kiliç, C. (2014). Posttraumatic growth after earthquake trauma is predicted by executive functions: A pilot study. *The Journal of Nervous and Mental Disease, 202*(12), 859–863. https://doi.org/10.1097/NMD.0000000000000211

Fuster, J. M. (2001). The prefrontal cortex – an update: Time is of the essence. *Neuron, 30*(2), 319–333. https://doi.org/10.1016/s0896-6273(01)00285-9

Guez, J., Naveh-Benjamin, M., Yankovsky, Y., Cohen, J., Shiber, A., & Shalev, H. (2011). Traumatic stress is linked to a deficit in associative episodic memory. *Journal of Traumatic Stress, 24*(3), 260–267. https://doi-org.ezproxy.bcu.ac.uk/10.1002/jts.20635

Harmon, J., & Venta, A. (2021). Adolescent posttraumatic growth: A review. *Child Psychiatry and Human Development, 52*(4), 596–608. https://doi-org.ezproxy.bcu .ac.uk/10.1007/s10578-020-01047-9

Hart, H., & Rubia, K. (2012). Neuroimaging of child abuse: A critical review. *Frontiers in Human Neuroscience, 6*, 52. https://doi.org/10.3389/fnhum.2012.00052

Hulbert, J. C., & Anderson, M. C. (2018). What doesn't kill you makes you stronger: Psychological trauma and its relationship to enhanced memory control. *Journal of Experimental Psychology: General, 147*(12), 1931–1949. https://doi.org/10.1037/ xge0000461

Hussain, D., & Bhushan, B. (2011). Posttraumatic stress and growth among Tibetan refugees: The mediating role of cognitive-emotional regulation strategies. *Journal of Clinical Psychology, 67*(4), 720–735. https://doi.org/10.1002/jclp.20801

Jin, Y., Xu, J., & Liu, D. (2014). The relationship between post traumatic stress disorder and post traumatic growth: Gender differences in PTG and PTSD sub-groups. *Social Psychiatry and Psychiatric Epidemiology, 49*, 1903–1910. https://doi .org/10.1007/s00127-014-0865-5

Kira, I. A., Shuweikh, H., Al-Huwailiah, A., El-wakeel, S. A., Waheep, N. N., Ebada, E. E., & Ibrahim, E. R. (2020). The direct and indirect impact of trauma types and cumulative stressors and traumas on executive functions. *Applied Neuropsychology: Adult, 29*(5). https://doi.org/10.1080/23279095.2020.1848835

Kolb, B. (1984). Functions of the frontal cortex of the rat: A comparative review. *Brain Research, 320*(1), 65–98. https://doi.org/10.1016/0165-0173(84)90018-3

McCrory, E., De Brito, S. A., & Viding, E. (2012). The link between child abuse and psychopathology: A review of neurobiological and genetic research. *Journal of the Royal Society of Medicine, 105*(4), 151–156. https://doi.org/10.1258/jrsm.2011.110222

Pechtel, P., & Pizzagalli, D. A. (2011). Effects of early life stress on cognitive and affective function: An integrated review of human literature. *Psychopharmacology, 214*, 55–70. https://doi.org/10.1007/s00213-010-2009-2

Perkins, N. H., & Meyers, A. (2020). The manifestation of physical and emotional sibling abuse across the lifespan and the need for social work intervention. *Journal of Family Social Work, 23*(4), 338–356. https://doi.org/10.1080/10522158 .2020.1799894

Scott, S. G., & Bruce, R. A. (1995). Decision-making style: The development and assessment of a new measure. *Educational and Psychological Measurement, 55*(5), 818–831. https://doi.org/10.1177/0013164495055005017

Stockton, H., Hunt, N. C., & Joseph, S. A. (2011). Cognitive processing, rumination, and posttraumatic growth. *Journal of Traumatic Stress, 24*(1), 85–92. https://doi .org/10.1002/jts.20606

Tedeschi, R. G., & Calhoun, L. G. (1996). The posttraumatic growth inventory: Measuring the positive legacy of trauma. *Journal of Traumatic Stress, 9*(3), 455–471. https://doi.org/10.1007/BF02103658

Tedeschi, R. G., & Calhoun, L. G. (2004). Posttraumatic growth: Conceptual foundations and empirical evidence. *Psychological Inquiry, 15*(1), 1–18. https://doi.org/ 10.1207/s15327965pli1501_01

Wilson, K. R., Hansen, D. J., & Li, M. (2011). The traumatic stress response in child maltreatment and resultant neuropsychological effects. *Aggression and Violent Behavior, 16*(2), 87–97. https://doi.org/10.1016/j.avb.2010.12.007

Wong, C. C. Y., & Yeung, N. C. Y. (2017). Self-compassion and posttraumatic growth: Cognitive processes as mediators. *Mindfulness, 8*(4), 1078–1087. https://doi.org/10.1007/s12671-017-0683-4

Young Minds. (2018). *Understanding trauma and adversity.* www.youngminds.org.uk/professional/resources/understanding-trauma-and-adversity

Zhang, Y., Xu, W., Yuan, G., & An, Y. (2018). The relationship between posttraumatic cognitive change, posttraumatic stress disorder, and posttraumatic growth among Chinese adolescents after the Yancheng Tornado: The mediating effect of rumination. *Frontiers in Psychology, 9*, 474. https://doi.org/10.3389/fpsyg.2018.00474

11

• • • • • • •

Decision-Making Deficits

11.1 What Is a Decision-Making Deficit?

A decision-making deficit by definition refers to an observable difference in the way an individual processes or reaches a decision. Depending on the circumstance, that could mean an individual is more prone to take risks, or perhaps that they are unable to make an informed choice.

Until this point, this book has focused on the underlying mechanisms and functionality of decision-making in a healthy brain, or neurotypical mind. Previous chapters have covered the networks and circuitry of the most complex cortical region, the prefrontal cortex (PFC). As evidence has shown, the PFC and associated functions, like decision-making, are highly complex. Along with the physical location of the PFC, this makes it a highly vulnerable region of the brain, in terms of being open to damage, as well as dysfunction in its complex development.

Dysfunction in the complex connections of the PFC leads to deficits in a range of cognitive functions and processes, just one of which is decision-making. Decision-making deficits are seen in various populations and vary greatly. For example, specific PFC dysfunction can lead to dramatic changes in cognition and action (Bast et al., 2017; Millan et al., 2012), which can be observed in cases such as Parkinson's disease, schizophrenia, attention deficit hyperactivity disorder, obsessive compulsive disorder, Tourette's syndrome, Huntington's disease and addiction (Callicott et al., 2003; Goldstein & Volkow, 2011; Joel, 2001; Rae et al., 2020; Williams &

Goldman-Rakic, 1998). This book does not have the scope to discuss all of these in detail; the focus will be on schizophrenia, autism, substance use, depression and antisocial personality disorder.

The PFC is often considered the seat of abstract thought and executive function, the CEO of the brain. This complex region dynamically interacts with multiple subcortical and other cortical areas, integrating information from all over the brain, so it is very highly connected to enable this. The complex PFC circuitry is dynamically modulated by neurotransmitters, specifically dopamine (Brozoski et al., 1979). Changes in this complex circuitry and how it behaves are thought to contribute to a range of disorders. Such changes may happen through development, or may be a result of trauma, disease or other damage. Notably, there are common features between seemingly very different psychological disorders, such as autism and schizophrenia, when it comes to decision-making.

11.2 How Do We Measure Decision-Making?

To gain a clear picture of decision-making deficits, or observable changes in decision-making, it must be possible to measure decision-making performance in some quantifiable way. Over time, many measures of decision-making have been developed, aimed at quantifying performance in a range of types of decision. One of the most commonly investigated types of decision is risky decision-making, for which a gambling style task is often employed.

11.3 Iowa Gambling Task

The Iowa gambling task (Bechara et al., 1994) replicates real-life decision-making in the context of risk and ambiguity, and helps researchers and practitioners determine the nature of impairment in decision-making. The Iowa gambling task involves choosing from four decks of cards, differing in terms of their reward–punishment profiles (Figure 11.1).

Repeated selection from two of the advantageous decks results in overall net profit, whilst repeated selection from the two disadvantageous decks results

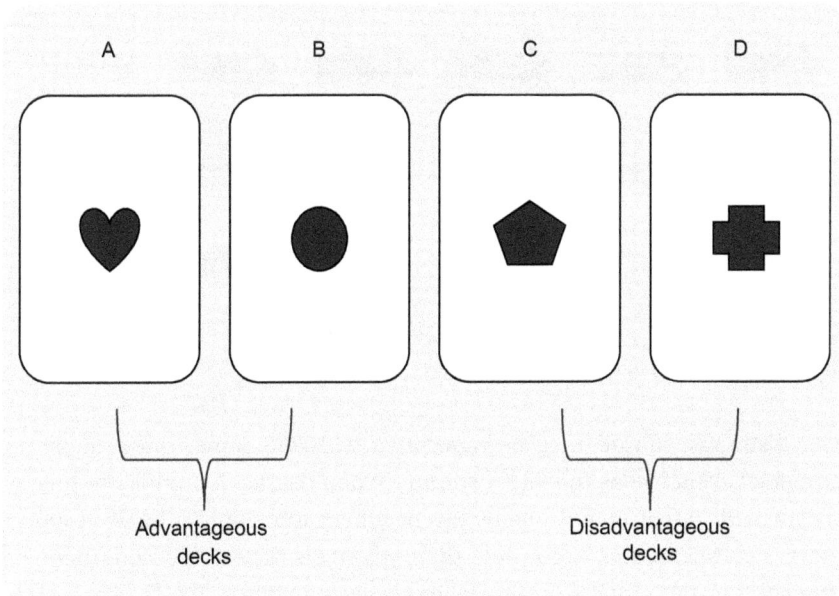

Figure 11.1 Example Iowa gambling task

in greater losses. The performance measure is the difference between the number of choices from the advantageous and disadvantageous decks, giving an overall 'net' score. The data collected through this task is intended to reflect the dynamic nature of the complex cognitive demands of risky decision-making.

Interestingly, patients with frontal lobe damage have consistently been shown to choose more from the disadvantageous decks (Brown et al., 2015), where there is always a potential for a higher win but a greater loss. This indicates greater risky decision-making, likely a result of dysfunctional connectivity associated with the decisions being made.

11.4 Flanker Task

The Eriksen flanker task (Eriksen & Eriksen, 1974) is another way of measuring an element of decision-making. It involves participants focusing on the centrally presented target stimuli, flanked on either side by distracter stimuli (Figure 11.2).

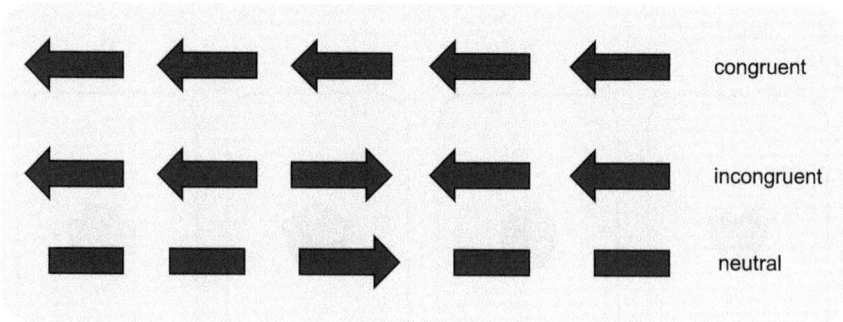

Figure 11.2 Example Eriksen flanker task

The distracter stimuli may be associated with the same, different or no particular response as the target stimuli. When distracter stimuli are incongruent with target stimuli, generally healthy participants reliably respond more slowly and less accurately than when distracters are congruent or unrelated to the target stimuli. Owing to these features, the flanker task is regarded as a classic test of cognitive control, referring to the ability to persist in goal-directed behaviour in the face of competing cognitive and behavioural demands.

11.5 Schizophrenia

To gain an understanding of how decision-making deficits play a role in schizophrenia, we must first clarify what characterises schizophrenia as a psychotic disorder. Schizophrenia is a neurodevelopmental disorder, that falls under the umbrella term of psychosis. It is characterised by the existence of positive and negative symptoms. Positive symptoms includes hallucinations and delusions. Negative symptoms include thought disorder and altered perception among others (American Psychiatric Association, 2013). Over the past few decades, research utilising brain imaging methods has established the presence of smaller frontal lobes in those with schizophrenia (Koeda et al., 2013), an abnormal density of neurons in the prefrontal region (Selemon et al., 1995) and reduced dendritic spine density in the prefrontal region (Glantz & Lewis, 2000). One researcher, Wible (Wible et al., 2001), measured prefrontal and medial–temporal cortical grey and white matter volume in chronic, male schizophrenic participants. Those with high negative symptom scores had significantly smaller bilateral white matter volumes than those with low negative symptom scores. These results underscore the importance of temporal–prefrontal pathways in the symptomatology of schizophrenia, especially where decision-making is concerned. These findings suggest an association between prefrontal connectivity and changes in decision-making, such as increased risky decision-making, as a result of the neuronal structural deficits.

Schizophrenia is often described as a disorder of neural connectivity. Dysfunction in the networks within the PFC has been associated with deficits in executive tasks, including decision-making (Stuss & Benson, 1984). Research findings conclude that dysfunctional neuronal connectivity attributes to risky and poor decision-making, especially in ambiguous contexts (Poudel et al., 2020).

Neuropathological studies of schizophrenia have identified changes in the way cells are organised in the PFC. Interestingly, evidence shows that these changes in organisation differ between first episode, chronic schizophrenics and healthy controls (Karlsgodt et al., 2010). These underlying structural and functional deficits give rise to the defining features of schizophrenia and the malfunction of the underlying neuronal networks that characterises psychosis, and subsequently negatively impacting decision-making processes.

Evidence from psychology research consistently shows people with schizophrenia often exhibit problems in being able to understand the relationship between their actions or behaviours and the outcomes of those behaviours, or the consequences (Morris et al., 2018; Voss et al., 2010). A result of this inability to make a link between action and outcome is a person who might keep selecting the same behaviours, despite an undesirable outcome. This is the case even if they know a different behaviour has a more valuable outcome. People suffering from schizophrenia are also often more willing than healthy counterparts to adopt strategies based on less information – in other words people with schizophrenia often jump to conclusions when making decisions.

The PFC has been significantly implicated in schizophrenia. The striatum is also involved – this is a structure known to be involved in the selection and initiation of a specific action in response to a decision, usually with a specific goal in mind (Balleine et al., 2007; Kim et al., 2009). As we know, the PFC helps us understand when a strategy is needed for successful changes.

Deficits involving both of these regions understandably have a significant impact on the way people make decisions.

If we think back to driving as an example again, in a healthy individual, if all traffic lights were suddenly replaced with sirens instead of lights to indicate we should stop, our PFC and striatum would both work to help us realise this change and adjust our decision-making process of stopping appropriately at junctions. When the anticipated outcome of a choice changes (such as if one was better, but then suddenly another option became better), the PFC helps us identify this.

Extensive research efforts across the disciplines of psychology and neuroscience have identified differences in brain function of schizophrenic patients. These differences have been observed in several regions but extensively so in the PFC. It is clear that changes in the PFC are a key element of schizophrenia. This evidence aligns with knowledge of schizophrenia symptoms linked to decision-making, and our existing knowledge of PFC importance in the decision-making process.

When it comes to the positive symptoms of schizophrenia, including hallucinations and delusions, there are known alterations in neurophysiology. The neurotransmitter dopamine is known to be involved in schizophrenia (Howes et al., 2009; Kegeles et al., 2010) and may be involved in the manifestation of both delusions and hallucinations. Dopamine is involved in the ability to anticipate rewards, controlling the physical actions necessary to act on choices, and in the decision-making process itself. A vast amount of research evidence has revealed increases in dopamine levels in the striatum (e.g. Boehme et al., 2015; Deserno et al., 2016; Egerton et al., 2013). This may be related to problems with the integration of information from elsewhere in the cortex, making informed decision-making difficult. The dopamine hypothesis of schizophrenia to this day is a key focus of aims to understand and treat the complex disorder.

DISCUSSION POINT

If disorders such as schizophrenia are understood to be partially a consequence of abnormal development in terms of decision-making networks, could an intervention in an important developmental stage prevent onset of the disorder later?

11.6 Substance Use

Substance use disorder, particularly with stimulants such as methamphetamine or cocaine, often leads to people reporting feelings of 'getting stuck' when certain outcomes change from the expected in a decision process. For example, if we reversed all the street lights so red meant 'go' and green meant 'stop' without telling anyone, most people would get an initial shock but would eventually alter their behaviour when they have got used to and understood the change. People with substance use disorder, specifically those dependent on methamphetamine or cocaine, however, would often take longer to learn to stop on the green light – even if they kept getting into collisions. This is because excessive stimulant use impacts regions in the brain that are crucial to adapting to changing environments – the striatum and the PFC. So these people are no longer able to adapt in the way they make this particular decision of when to stop and when to go; they will continue to make the same decision with the poorer outcome, much like we saw earlier in patients with schizophrenia.

Stimulants also affect neurotransmission in the areas of the brain involved in decisions. They cause excessive dopamine release, which can alter the balance between goal-directed behaviours and habits. Goal-directed behaviours are behaviours which are flexible and respond to environmental changes. Habits are automatic and hard to break. Usually, when we learn something new our brain keeps adapting and incorporating new information. But this is slow and cognitively demanding. Substance dependence can accelerate a person's progression to habitual behaviour, wherein a set strategy or response become ingrained. Not only does this impact behaviours, it makes it hard to stop seeking the drugs themselves, even if the individual no longer finds the experience of taking them enjoyable.

11.7 Autism

Making decisions can present significant problems for individuals with autism. Research evidence has suggested differences between the decision-making of adults with autism diagnoses and their neurotypical counterparts (Luke et al., 2012). Research has sought to investigate this further, by comparing the real-life decision-making experiences of adults with and without autism diagnoses (Luke et al., 2012). The researchers hypothesised that compared with a neurotypical group, participants with autism would report more frequent experiences of problems during decision-making, such as finding the process tiring. They also expected to see people with autism to show greater difficulty with particular features of decisions, such as decisions that needed to be made quickly. The experimenters also expected to find greater reliance on rational, avoidant and dependent styles of decision-making. The participants in the study completed a questionnaire to evaluate their decision-making experiences. The questionnaire asked participants to rate the frequency with which particular problems in decision-making were experienced, the extent to which they perceived difficulties in relation to particular features of decisions and finally the extent to which they believed that their condition of autism enhanced or interfered with their own decision-making. Participants also completed the General Decision-Making Style Questionnaire. Levels of anxiety and depression were assessed using the well-established Hospital Anxiety and Depression Scale. The results of this study indicated that compared with their neurotypical peers, people with autism more frequently report difficulties in making decisions. Decisions that needed to be made quickly or involved a change of routine or talking to others were experienced as particularly difficult, and the process of decision-making was reported to be exhausting, overwhelming and anxiety-provoking by individuals with autism. Those with autism reported significantly higher levels of anxiety and depression in comparison to neurotypical counterparts and were more likely to believe that their having autism interfered with rather than enhanced their decision-making process. Interestingly, people with autism are more likely to say they avoid decision-making.

The overall findings of the study suggest that, compared with neurotypical individuals, people with autism experience greater difficulty with decision-making. Decision-making in those with autism was associated with anxiety, exhaustion, problems engaging in the decision-making process, and a tendency to avoid decision-making. These findings are consistent with previous individual accounts, known characteristics of autism and previous studies of decision-making in autism. It is important to consider that the difficulties reported by those with autism could be further exacerbated by the higher levels of anxiety and depression found in that group. It is widely established that anxiety and depression are also associated with differences in decision-making, so this could have an impact.

The researchers found that ratings of perceived frequency of interference from autism increased along with levels of anxiety and depression. The results are consistent with suggestions from the literature relating to decision-making for people with autism. Importantly, they also have some practical implications for supporting adults with autism. For example, it may be useful to provide additional time to reach a choice, minimise irrelevant information, present closed questions, offer encouragement and reassurance and address general issues around anxiety. Understanding how adults with autism experience decision-making is especially relevant for family members and professionals who are involved in providing support to help these individuals achieve greater self-understanding, self-advocacy and improved decision-making in areas such as employment.

Other research evidence also indicates differences in the way people with autism make decisions compared to a neurotypical population. For example, Minassian et al. (2017) found that unlike the control group, people with autism demonstrate a more pronounced 'win–stay/lose–shift' strategy when the error rate is low. This suggests that people with autism may be influenced to a greater extent by reinforcement. Similarly, Damiano et al. (2014) stated that adults with autism are prepared to expend more effort for monetary rewards than control participants but demonstrate reduced sensitivity to reward. The authors related this to the high levels, among people with autism spectrum disorders, of circum-scribed interests, often pursued at any cost. More broadly, it suggests that, in some contexts, people with autism may be less flexible in their decision-making.

11.8 Depression

Major depressive disorder is often characterised by two main symptoms: sustained negative affect and reduced positive affect. In general, decisions carried out when a patient is in a depressed state are affected by negative affect and often distorted negative cognitions. A diagnosis of depression often includes symptoms of low self-worth, lack of drive and so on. These symptoms are said to directly impact decisions made. Although, interestingly, some research evidence does suggest that mild depressive symptoms could be related with more realistic self-assessments. This phenomenon is called depressive realism.

11.9 Depressive Realism

Depressive realism is a concept that was first identified by Alloy and Abramson (1980) and is based on the idea that depressed individuals make more realistic inferences and decisions than non-depressed people. Their research shows that negative cognitive bias associated with depression reflects accuracy. Whilst considering how this idea applies to our understanding of decision-making deficit in depression, it is important to acknowledge that some scholars have argued that the evidence presented to support depressive realism as a concept does not apply beyond laboratory settings and cannot be applicable to real-life settings.

Alloy and Abramson concluded that depressed individuals believe they have low causal agency – that is, they believe that they are not in control of events around them. This appears to fit with the known symptoms and diagnoses. It is plausible that depression is linked to an external locus of control. A locus of control is the degree to which people believe that they, as opposed to external forces, have control over the outcome of events in their lives.

So is being a depressive realist advantageous when it comes to decisions? It certainly sounds it. We know that having depression is not considered an advantage, and does not leave those suffering feeling positive, so this conclusion is understandably a little confusing. The analytical rumination hypothesis (Andrews & Thomson, 2009) suggests depression is an advantageous trait. Evidence shows that response to complex problems minimises disruption of rumination in depression. This is very much in contrast with the common view that depression is maladaptive, which there is a lot of evidence to support. The analytical rumination hypothesis claims depressive rumination helps people solve problems and make decisions.

11.10 Differences in Decision-Making in Depression

A large body of research evidence shows a consistent pattern of reduced activation of the ventral striatum, dorsal striatum and PFC in response to positive reward stimuli in decision-making scenarios (Arrondo et al., 2015). A real-life consequence of altered reward processing in depression has been demonstrated using an ultimatum game in which participants were asked to accept or reject a wide range of offers (Peterburs et al., 2017). Depressed patients typically exhibit a more negative emotional reaction to unfair offers, despite accepting more of these offers than controls. Thus, depression appears to have a dual effect on the processing of reward and value: induction of excessive emotional responses and reduced willingness to reject unfair offers.

Like other disorders associated with decision-making deficit, there appears to be an alteration in the dopaminergic and serotonergic systems in people with diagnosed depression. Specifically, dopamine neurotransmission in major depressive disorder seems to be reduced (Bekhbat et al., 2022). It is also known that selective serotonin reuptake inhibitors (SSRIs) work by acting on the availability of serotonin, so it is clear serotonin plays a role. We know this because of the successful use of SSRIs and also through studies that show serotonin depletion through diet can induce symptoms of depression.

Anhedonia, which is the loss of pleasure or interest in previously rewarding stimuli, is a key pathological element of depression and predicts antidepressant response. Anhedonia is associated with disruption of the frontostriatal valuation system and reward processing (Steinmann et al., 2022) – that is, a disruption to one of the many systems or circuits involving the PFC and striatum. Anhedonic depressed patients also often exhibit a physically reduced volume of specific parts of the PFC, which could lead to reduced connectivity. You will begin to see that many of the causes of decision-making deficit seem to stem down to changes in the connectivity of the PFC.

Suicide is the most catastrophic outcome of depression and is associated with reduced serotonergic neurotransmission, particularly within the PFC. This is thought to impair executive function, predisposing patients to become more impulsive, rigid in their thinking and poorer decision-makers.

Deficits in executive function and problem-solving are greater in depressed individuals with a history of suicide attempts or even suicidal ideation compared with depressed controls (Fernández-Sevillano et al., 2021).

11.11 Antisocial Personality Disorder

Psychopathy is a chronic and persistent developmental personality disorder characterised by emotional dysfunction, displaying shallow affect, callous unemotional traits, antisocial behaviours, impulsivity, risk-taking, sensation-seeking and interpersonal manifestations of grandiosity, egocentricity and manipulation. Psychopathy is not recognised as a disorder by the DSM-5, rather we refer to antisocial personality disorder, which encompasses the psychopathic traits or symptoms often described in psychopathy.

Research has yet to establish a clear relationship between psychopathic traits and neurocognitive dysfunctional connectivity involved in executive functioning and decision-making. However, common patterns of atypical connectivity within the PFC have been observed, and these are thought to be a possible contributor to symptoms related to poorer cognitive control and decision-making. Through lesion studies (Bechara et al., 1994), it has been established that damage to the amygdala and PFC lead to risky decisions that are not guided by consideration of future consequences. Disruptions in the integration of decision-making neural circuits may therefore impact on the processing of information, impacting reward and punishment, consequently leading to poorer decision-making. It is important to note that these key symptoms do not solely manifest in those we might describe as psychopaths, but are also widely prevalent in those with psychopathic traits amongst a general population.

A systematic review (Hughes et al., 2015) explored psychopathology and decision-making and concluded that further understanding of psychopathy and neurocognitive functions, including further research on decision-making, could have important implications for diagnosis, refinement of the characterisation of psychopathic traits, increase understanding of aetiological factors and discover which traits group together from a neurocognitive perspective.

11.12 Can We Improve Decision-Making Deficits?

There are undisputed benefits to some pharmacological treatments such as SSRIs for depression, that can aid in combatting some of the difficulties identified with decision-making. Unfortunately, however, SSRIs do not work for everyone, and cognitive deficits like decision-making are difficult to treat. There are no medications for schizophrenia or stimulant dependence shown to reliably improve decision-making. This is a consequence of the human brain's complexity, that modern science has not yet matched.

Although there is no quick fix, we can all improve our memory and decision-making through certain activities. For instance, cognitive remediation therapy is a behavioural approach that trains the brain to respond to certain situations better. For people with schizophrenia, it may improve visual memory and perhaps more complex decision-making.

Not being able to navigate decisions day-to-day is one of the most debilitating aspects of disorders that impact cognition. This leads to difficulties in maintaining work, keeping friends and leading a fulfilling life. Further research is vital to fully understand the complexities of human decision-making, and how different people make different decisions. Only through gaining a complete understanding of decision-making in a healthy brain can we begin to build more effective interventions for those living with disorders associated with decision-making deficit.

Key Points

- Anatomical differences in the prefrontal cortex are seen in schizophrenia, which is thought to impact decision-making.
- People with autism report more frequent experiences of problems during decision-making, specifically when decisions need to be made quickly or involve a change in routine.
- The concept of depressive realism tells us that people with depression make more realistic decisions than healthy comparisons.

REFERENCES

Alloy, L. B., & Abramson, L. Y. (1980). Judgment of contingency in depressed and nondepressed students: Sadder but wiser? *Journal of Experimental Psychology, 108*(4), 441–485.

American Psychiatric Association. (2013). *Diagnostic and statistical manual of mental disorders: DSM-5* (5th ed.). American Psychiatric Publishing, Inc. https://doi.org/10.1176/appi.books.9780890425596

Andrews, P. W., & Thomson, J. A. (2009). The bright side of being blue: Depression as an adaptation for analyzing complex problems. *Psychological Review, 116*, 620–654. https://doi.org/10.1037/a0016242

Arrondo, G., Segarra, N., Metastasio, A., Ziauddeen, H., Spencer, J., Reinders, N. R., Dudas, R. B., Robbins, T. W., Fletcher, P. C., & Murray, G. K. (2015). Reduction in ventral striatal activity when anticipating a reward in depression and schizophrenia: A replicated cross-diagnostic finding. *Frontiers in Psychology, 6*, 1280. https://doi.org/10.3389/fpsyg.2015.01280

Balleine, B. W., Delgado, M. R., & Hikosaka, O. (2007). The role of the dorsal striatum in reward and decision-making. *Journal of Neuroscience, 27*(31), 8161–8165. https://doi.org/10.1523/JNEUROSCI.1554-07.2007

Bast, T., Pezze, M., & McGarrity, S. (2017). Cognitive deficits caused by prefrontal cortical and hippocampal neural disinhibition. *British Journal of Pharmacology, 174*(19), 3211–3225. https://doi.org/10.1111/bph.13850

Bechara, A., Damasio, A. R., Damasio, H., & Anderson, S. W. (1994). Insensitivity to future consequences following damage to human prefrontal cortex. *Cognition, 50*(1–3), 7–15. https://doi.org/10.1016/0010-0277(94)90018-3

Bekhbat, M., Li, Z., Mehta, N. D., Treadway, M. T., Lucido, M. J., Woolwine, B. J., Haroon, E., Miller, A. H., & Felger, J. C. (2022). Correction to: Functional connectivity in reward circuitry and symptoms of anhedonia as therapeutic targets in depression with high inflammation: Evidence from a dopamine challenge study. *Molecular Psychiatry, 27*(10), 4122. https://doi.org/10.1038/s41380-022-01754-w

Boehme, R., Deserno, L., Gleich, T., Katthagen, T., Pankow, A., Behr, J., Buchert, R., Roiser, J. P., Heinz, A., & Schlagenhauf, F. (2015). Aberrant salience is related to reduced reinforcement learning signals and elevated dopamine synthesis capacity in healthy adults. *Journal of Neuroscience, 35*(28), 10103–10111. https://doi.org/10.1523/JNEUROSCI.0805-15.2015

Brown, E. C., Hack, S. M., Gold, J. M., Carpenter Jr., W. T., Fischer, B. A., Prentice, K. P., & Waltz, J. A. (2015). Integrating frequency and magnitude information in decision-making in schizophrenia: An account of patient performance on the Iowa Gambling Task. *Journal of Psychiatric Research, 66–67*, 16–23. https://doi.org/10.1016/j.jpsychires.2015.04.007

Brozoski, T. J., Brown, R. M., Rosvold, H. E., & Goldman, P. S. (1979). Cognitive deficit caused by regional depletion of dopamine in prefrontal cortex of rhesus monkey. *Science, 205*(4409), 929–932. https://doi.org/10.1126/science.112679

Callicott, J. H., Mattay, V. S., Verchinski, B. A., Marenco, S., Egan, M. F., & Weinberger, D. R. (2003). Complexity of prefrontal cortical dysfunction in schizophrenia: More than up or down. *The American Journal of Psychiatry*, *160*(12). https://doi.org/10.1176/appi.ajp.160.12.2209

Damiano, C. R., Mazefsky, C. A., White, S. W., & Dichter, G. S. (2014). Future directions for research in autism spectrum disorders. *Journal of Clinical Child & Adolescent Psychology*, *43*(5), 828–843. https://doi.org/10.1080/15374416.2014.945214

Deserno, L., Schlagenhauf, F., & Heinz, A. (2016). Striatal dopamine, reward, and decision making in schizophrenia. *Dialogues in Clinical Neuroscience*, *18*(1), 77–89. https://doi.org/10.31887/DCNS.2016.18.1/ldeserno

Egerton, A., Chaddock, C. A., Winton-Brown, T. T., Bloomfield, M. A. P., Bhattacharyya, S., Allen, P., McGuire, P. K., & Howes, O. D. (2013). Presynaptic striatal dopamine dysfunction in people at ultra-high risk for psychosis: Findings in a second cohort. *Biological Psychiatry*, *74*(2), 106–112. https://doi.org/10.1016/j.biopsych.2012.11.017

Eriksen, B. A., & Eriksen, C. W. (1974). Effects of noise letters upon the identification of a target letter in a nonsearch task. *Perception & Psychophysics*, *16*(1), 143–149. https://doi.org/10.3758/BF03203267

Fernández-Sevillano, J., Alberich, S., Zorrilla, I., González-Ortega, I., López, M. P., Pérez, V., Vieta, E., González-Pinto, A., & Saíz, P. (2021). Cognition in recent suicide attempts: Altered executive function. *Frontiers in Psychiatry*, *12*, 701140. https://doi.org/10.3389/fpsyt.2021.701140

Glantz, L. A., & Lewis, D. A. (2000). Decreased dendritic spine density on prefrontal cortical pyramidal neurons in schizophrenia. *Archives of General Psychiatry*, *57*(1), 65–73. https://doi.org/10.1001/archpsyc.57.1.65

Goldstein, R. Z., & Volkow, N. D. (2011). Dysfunction of the prefrontal cortex in addiction: Neuroimaging findings and clinical implications. *Nature Reviews Neuroscience*, *12*(11), 652–669. https://doi.org/10.1038/nrn3119

Howes, O. D., Montgomery, A. J., Asselin, M.-C., Murray, R. M., Valli, I., Tabraham, P., Bramon-Bosch, E., Valmaggia, L., Johns, L., Broome, M., McGuire, P. K., & Grasby, P. M. (2009). Elevated striatal dopamine function linked to prodromal signs of schizophrenia. *Archives of General Psychiatry*, *66*(1), 13–20. https://doi.org/10.1001/archgenpsychiatry.2008.514

Hughes, M. A., Dolan, M. C., & Stout, J. C. (2015). Decision-making in psychopathy. *Psychiatry, Psychology and Law*, *23*(4), 1–17. https://doi.org/10.1080/13218719.2015.1081228

Joel, D. (2001). Open interconnected model of basal ganglia-thalamocortical circuitry and its relevance to the clinical syndrome of Huntington's disease. *Movement Disorders*, *16*(3), 407–423. https://doi.org/10.1002/mds.1096

Karlsgodt, K. H., Sun, D., & Cannon, T. D. (2010). Structural and functional brain abnormalities in schizophrenia. *Current Directions in Psychological Science*, *19*(4), 226–231. https://doi.org/10.1177/0963721410377601

Kegeles, L. S., Abi-Dargham, A., Frankle, W. G., Gil, R., Cooper, T. B., Slifstein, M., Hwang, D.-R., Huang, Y., Haber, S. N., & Laruelle, M. (2010). Increased synaptic dopamine function in associative regions of the striatum in schizophrenia. *Archives of General Psychiatry, 67*(3), 231–239. https://doi.org/10.1001/archgenpsychiatry.2010.10

Kim, H., Sul, J. H., Huh, N., Lee, D., & Jung, M. W. (2009). Role of striatum in updating values of chosen actions. *Journal of Neuroscience, 29*(47), 14701–14712. https://doi.org/10.1523/JNEUROSCI.2728-09.2009

Koeda, M., Takahashi, H., Matsuura, M., Asai, K., & Okubo, Y. (2013). Cerebral responses to vocal attractiveness and auditory hallucinations in schizophrenia: A functional MRI study. *Frontiers in Human Neuroscience, 7,* 221. https://doi.org/10.3389/fnhum.2013.00221

Luke, L., Clare, I. C. H., Ring, H., Redley, M., & Watson, P. (2012). Decision-making difficulties experienced by adults with autism spectrum conditions. *Autism, 16*(6), 612–621. https://doi.org/10.1177/1362361311415876

Millan, M. J., Agid, Y., Brune, M., Bullmore, E. T., Carter, C. S., Clayton, N. S., Connor, R., Davis, S., Deakin, B., DeRubeis, R. J., Dubois, B., Geyer, M. A., Goodwin, G. M., Gorwood, P., Jay, T. M., Joëls, M., Mansuy, I. M., Meyer-Lindenberg, A., Murphy, D., ... Young, L. J. (2012). Cognitive dysfunction in psychiatric disorders: Characteristics, causes and the quest for improved therapy. *Nature Reviews Drug Discovery, 11,* 141–168. https://doi.org/10.1038/nrd3628

Minassian, K., Hofstoetter, U. S., Dzeladini, F., Guertin, P. A., & Ijspeert, A. (2017). The human central pattern generator for locomotion: Does it exist and contribute to walking? *The Neuroscientist, 23*(6), 649–663. https://doi.org/10.1177/1073858417699790

Morris, R. W., Cyrzon, C., Green, M. J., Le Pelley, M. E., & Balleine, B. W. (2018). Impairments in action-outcome learning in schizophrenia. *Translational Psychiatry, 8*(1), 54. https://doi.org/10.1038/s41398-018-0103-0

Peterburs, J., Voegler, R., Liepelt, R., Schulze, A., Wilhelm, S., Ocklenburg, S., & Straube, T. (2017). Processing of fair and unfair offers in the ultimatum game under social observation. *Scientific Reports, 7,* 44062. https://doi.org/10.1038/srep44062

Poudel, R., Riedel, M. C., Salo, T., Flannery, J. S., Hill-Bowen, L. D., Eickhoff, S. B., Laird, A. R., & Sutherland, M. T. (2020). Common and distinct brain activity associated with risky and ambiguous decision-making. *Drug and Alcohol Dependence, 209,* 107884. https://doi.org/10.1016/j.drugalcdep.2020.107884

Rae, C. L., Parkinson, J., Betka, S., Gouldvan Praag, C. D., Bouyagoub, S., Polyanska, L., Larsson, D. E. O., Harrison, N. A., Garfinkel, S. N., & Critchley, H. D. (2020). Amplified engagement of prefrontal cortex during control of voluntary action in Tourette syndrome. *Brain Communications, 2*(2), fcaa199. https://doi.org/10.1093/braincomms/fcaa199

Selemon, L. D., Rajkowska, G., & Goldman-Rakic, P. S. (1995). Abnormally high neuronal density in the schizophrenic cortex: A morphometric analysis of prefrontal area 9 and occipital area 17. *Archives of General Psychiatry, 52*(10), 805–820. https://doi.org/10.1001/archpsyc.1995.03950220015005

Steinmann, L. A., Dohm, K., Goltermann, J., Richter, M., Enneking, V., Lippitz, M., Repple, J., Mauritz, M., Dannlowski, U., & Opel, N. (2022). Understanding the neurobiological basis of anhedonia in major depressive disorder: Evidence for reduced neural activation during reward and loss processing. *Journal of Psychiatry and Neuroscience, 47*(4), E284–E292. https://doi.org/10.1503/jpn.210180

Stuss, D. T., & Benson, D. F. (1984). Neuropsychological studies of the frontal lobes. *Psychological Bulletin, 95*(1), 3–28. https://doi.org/10.1037/0033-2909.95.1.3

Voss, M., Moore, J., Hauser, M., Gallinat, J., Heinz, A., & Haggard, P. (2010). Altered awareness of action in schizophrenia: A specific deficit in predicting action consequences. *Brain, 133*(10), 3104–3112. https://doi.org/10.1093/brain/awq152

Wible, C. G., Anderson, J., Shenton, M. E., Kricun, A., Hirayasu, Y., Tanaka, S., Levitt, J. J., O'Donnell, B. F., Kikinis, R., Jolesz, F. A., & McCarley, R. W. (2001). Prefrontal cortex, negative symptoms, and schizophrenia: An MRI study. *Psychiatry Research, 108*(2), 65–78. https://doi.org/10.1016/S0925-4927(01)00109-3

Williams, S. M., & Goldman-Rakic, P. S. (1998). Widespread origin of the primate mesofrontal dopamine system. *Cerebral Cortex, 8*(4), 321–345. https://doi.org/10.1093/cercor/8.4.321

12

• • • • • • •

Implications of Decision-Making

12.1 Consumer Behaviour

12.1.1 Influencing Decisions: Cognitive Bias

Cognitive bias is an umbrella term used to describe a systematic pattern in thought processes, often based on error or unconscious judgement. These patterns are often automatic. Cognitive bias may include confirmation bias, negativity bias, halo effect, availability heuristic, attentional bias, optimism bias, bandwagon effect, loss aversion and ingroup bias amongst others.

For the purpose of understanding the role of cognitive bias in complex decision-making, we need to recall the importance of efficiency of processing when making complex decisions. As established through earlier chapters, the decision process can involve vast networks and the integration of information from multiple pathways and systems. It is advantageous, therefore, that the brain does what it can to be as efficient in this as possible. Systematic behaviours, or automatic, learned processes are helpful in this respect. But, the brain is after all a living thing, so is not immune to errors. The systematic behaviours that are cognitive biases can sometimes lead an individual to an ill-informed decision, or to making a decision based on subjective preconceived ideas rather than objective evidence. This is a bit like acting on a prejudice. Sometimes, decisions made based on cognitive biases, although unconscious, could lead to illogical decisions.

Confirmation bias is a good example of how cognitive bias influences complex decision-making. Confirmation bias is the process of an individual seeking out evidence to support existing beliefs, and ignoring evidence that contradicts it. This would be like writing an essay on the nature versus nurture debate based only on evidence supporting the nurture side of the argument, because you believe that to be true. This type of bias often works to the advantage of advertisers, because consumers following confirmation bias may only seek evidence to support the benefits of the advertised product. Advertisers and marketers often maximise on confirmation bias effects by researching the general views and beliefs of their target population and tailoring their advertising to confirm these viewpoints. For example, a shampoo company might target a group they know to be interested in the environment by advertising plastic-free packaging. The idea is that this confirmation of beliefs will result in the target audience aligning themselves with the brand as having common beliefs.

Bagchi et al. (2020) investigated the effects of consumers' confirmation bias on advertising strategies. Interestingly, their findings revealed the utilisation of confirmation bias does not improve profits in the short term but had a much greater effect in the long term. This suggests consumers are essentially building a long-term relationship with the products they purchase, when confirmation bias is involved in their decision process.

12.1.2 Confirmation Bias in a Social Media World

In the modern, internet-led world, we have immediate access to an abundance of information. In the past, we may have received our local and world news via a newspaper, or TV news, where we had no control or choice over the content we were presented with. In modern times, we are able to select which news items and topics we are exposed to. This is called selective exposure. The same applies to engagement with social media. For instance, if a person only follows cat accounts on popular social media sites, they are only going to see content from those cat accounts and won't see the content they do not select to be exposed to. In modern society, we are able to compose our own media diet (Van der Meer et al., 2020). Van der Meer et al. (2020) demonstrated that confirmation bias has a significant impact on the selective exposure to news on specific topics including immigration and privatisation of healthcare.

12.1.3 What Is the Function of Confirmation Bias?

We did not evolve this phenomenon of cognitive bias to aid advertisers in their long-term goals of selling products to us. Advertising and marketing did not exist when the human brain and its complexities in high-order decision-making evolved. So what is it for? Given its rather unhelpful nature, in causing us to make illogical decisions, it is not immediately clear why we would have evolved cognitive biases at all. Many have argued the process must be adaptive in some way (e.g. Rollwage & Fleming, 2021), that perhaps it saves time and enables the efficiency needed for complex processes. Others (e.g. Peters, 2022) suggest that confirmation bias came about for social advancement – its existence helps us influence others and their beliefs in a social context. The idea here is that we are social beings and to progress we need to be part of a social group with similar beliefs and ideas.

12.1.4 The Psychology of Advertising

In modern society, we are constantly presented with attempts to sway our emotions and influence our decisions, in the form of advertising. We see or hear adverts all day through the radio, TV, social media, emails, on public transport, on billboards and in magazines. They are all trying to influence our decision-making, to buy the product or service.

There are several ways advertisers aim to influence our decisions, which I will address next.

12.1.5 Persuasion and Manipulation

Persuasive advertising is a somewhat controversial topic. The term refers to a strategy of advertising that aims to convince target audiences of a need or desire for the advertised product or service. Many such persuasive adverts highlight the benefits of the product, for example for health, well-being or convenience.

12.1.6 Memory and Positive Associations

Some advertisers utilise association to aid us in building positive associations with their product. This may be by tapping into a pleasant memory, such as something nostalgic, or a popular song that elicits a positive emotive response. You will then associate the same positive feeling with the advertised product, making you more likely to buy it. Adverts engage celebrity endorsers in the same way. If you already have a positive association with, say, David Beckham, you are more likely to make the decision to buy the aftershave he is seen with in the advert. You might have noticed that when celebrities are suddenly part of a media scandal, they are quickly dropped from advertising campaigns, because now they will produce a negative association!

12.1.7 The Power of Colour

Colour is a useful psychological tool. You may have heard of warm and cool colours. Advertising takes this notion one step further, using psychology. Different colours can elicit different emotional responses. For example, blue is known to be calming and red is associated with agitation. Research evidence shows that the careful use of colour in advertising can influence perceptions of the advertised product and can also create brand recognition (Cunningham, 2017) – most people would immediately recognise Tiffany blue!

12.2 Free Will

The question of whether we as humans have free will in our actions and behaviours has long been a debate, across psychology, neuroscience, social sciences and philosophy. At the heart of the debate is the question of whether our actions are determined by external factors or if we have complete autonomy over our decision-making. When exploring how and why we make certain decisions, naturally we arrive at this time-old question, as to whether we have complete conscious control over the decisions we make. We know from the topics covered in earlier chapters of this book that, in fact, many decisions we make we are indeed making subconsciously or automatically. But, we have also established that we do have the ability to stop this automatic behaviour where there is a change of circumstances, so does this mean we have complete free will?

12.2.1 The Conscious and Subconscious of Decision-Making: Cognitive Experiential Theory of Decision-Making

Epstein (1973) developed the cognitive experiential theory (CET) of decision-making in an attempt to answer the age-old question as to whether we truly possess free will. The theory states that decision-making is controlled by two distinct systems, that function simultaneously but entirely independent of one another.

1. Rational system: conscious decision-making, uses cognitive processes such as problem-solving and reasoning.
2. Experiential system: may include unconscious decision-making, uses gut feelings, immediate desires and emotions.

This model moves a little away from traditional ideas of free will existing or not, on a binary level, and more towards an idea that there is an interaction between elements of conscious and unconscious, resulting in somewhat of a combination of decisions being based on free will and not. That is, due to the simultaneous processing of the two systems, each decision has both conscious and unconscious elements. For instance, one decision may have some rational input based on problem-solving but also contain some emotional input.

12.2.2 Contention Scheduling and Supervisory Attention System

Contention scheduling refers to a cognitive mechanism proposed by Norman and Shallice (1980, 1986). The concept of contention scheduling aims to offer an explanation for how we overcome competing cognitive processes. The human brain is complex, and there are often simultaneously competing, demanding, cognitive processes. To carry out multiple processes at the same time, to a high level, would be too cognitively demanding and require too much energy to be viable. There are limited cognitive resources, and it must be determined which of the competing cognitive processes at a given time will gain the resources and therefore be prioritised in terms of action. This process is called contention resolution, where the most relevant or urgent process is executed as a priority. Other competing tasks are inhibited. The idea of contention scheduling relies on the assumption that there is a finite amount of cognitive resources to be allocated.

When we consider contention scheduling in the context of free will, the theory implies we will not always have free will, specifically when it comes to the allocation of cognitive resources. We may not always have conscious input as to which competing cognitive process will take priority.

Following from the idea of contention scheduling, Shallice (1988) devised the concept of a supervisory attention system. This is described as a high-order cognitive mechanism with the purpose of regulating and co-ordinating lower-level cognitive processes – an executive control system. It is said that this system allows the management and prioritisation of competing cognitive processes. This system allows for the mediation of various behaviours, including automatic and habitual behaviours as well as more conscious, goal-directed actions. The extent to which we are able to consciously control the prioritisation of automatic versus conscious behaviours relates to how much free will we might have.

12.2.3 Hierarchical Structure of Decision-Making

Stuss and Benson (1986) proposed the hierarchical model of decision-making, encompassing three levels of cognitive control:

1. The strategic level allows for goal direction and future planning, based on beliefs and values.
2. The tactical level allows the selection and coordination of strategies to achieve goals.
3. The executional level allows the implementation and execution of strategies through actions.

An important element of this hierarchical structure is that each level follows the previous; they do not function in isolation, like other models of decision-making suggest. Each level allows for processing at increasing levels of abstraction from the bottom to the top of the hierarchy.

In response to the question as to whether we have free will, the hierarchical model of decision-making provides a more complex answer. At the higher, more abstract level of the hierarchy, an individual is likely to have free will – to have a perceived conscious control over their actions. However, at the lower levels of the hierarchy there is likely much less conscious control, for example in automatic responses and emotional reactions.

12.2.4 The Neurobiology of Free Will

If we consider the underlying neurobiology, we may be led to conclude that we do not technically have free will, as evidence shows neurophysiological processes take place before we are consciously aware of the decision (Guggisberg & Mottaz, 2013). Based solely on the biological evidence, we do not have free will. Neuroscientist Robert Sapolsky (Revell, 2013) claims that if you ask why a behaviour happened, you are essentially asking what triggered this response. On a biological level, that can be translated to what neurophysiological activity caused this particular series of responses? If you think on a grander scale, you could be asking what previous experiences, environmental influencers and genetic information contributed to neuro-transmission occurring in this manner? This could go on as far as consider-ing the actions of distant ancestors and how they ultimately influenced the decision of an individual today. If we consider all of these factors, out of our control, that influence how our brain is structured and connected, we do not have control over these things, but they do influence how we make a decision – so in that sense, we have no free will.

12.2.5 Do We Really Have Free Will?

As a passing thought, it may seem obvious that we as humans have free will. We have the capacity to make our own decisions. But if we carefully consider all of the contributing factors to the way in which we make deci-sions – genetics, environmental influences, childhood experiences, develop-mental factors, cultural influence, past experiences and memory and the neurophysiological process that underpins a decision, it becomes clear that every decision is not as controlled by our conscious thought as we might have first assumed. Ultimately, the concept of free will is a philosophical one, and not something science can expect to address.

Key Points

- Cognitive bias is a systematic, unconscious thought process based on judgement.
- The evolutionary advantage of cognitive bias is not clear, but some argue it is time saving and enables social advancement.
- Contention scheduling implies we will not always have free will, specifically when it comes to the allocation of cognitive resources.

REFERENCES

Bagchi, R., Ham, S. H., & He, C. (2020). Strategic implications of confirmation bias-inducing advertising. *Production and Operations Management, 29*(6), 1573–1596. https://doi.org/10.1111/poms.13176

Cunningham, M. (2017). The value of color research in brand strategy. *Open Journal of Social Sciences, 5*, 186–196. https://doi.org/10.4236/jss.2017.512014

Epstein, S. (1973). The self-concept revisited: Or a theory of a theory. *American Psychologist, 28*(5), 404–416. http://doi.org/10.1037/h0034679

Guggisberg, A. G., & Mottaz, A. (2013). Timing and awareness of movement decisions: Does consciousness really come too late? *Frontiers in Human Neuroscience, 7*, 385. https://doi.org/10.3389/fnhum.2013.00385

Norman, D. A., & Shallice, T. (1980). *Attention to action: Willed and automatic control of behavior.* CHIP Report 99, University of California, San Diego.

Norman, D. A., & Shallice, T. (1986). Attention to action: Willed and automatic control of behavior. In R. Davidson, G. Schwartz, & D. Shapiro (Eds.), *Consciousness and self-regulation: Advances in research and theory* (Vol. 4, pp. 1–18). Plenum.

Peters, U. (2022). What is the function of confirmation bias?. *Erkenntnis, 87*, 1351–1376. https://doi.org/10.1007/s10670-020-00252-1

Revell, T. (2013, 18 October). Why free will doesn't exist, according to Robert Sapolsky. *New Scientist.* https://tinyurl.com/jhxf8xu5

Rollwage, M., & Fleming, S. M. (2021). Confirmation bias is adaptive when coupled with efficient metacognition. *Philosophical Transactions of the Royal Society B: Biological Sciences, 376*, 20200131. https://doi.org/10.1098/rstb.2020.0131

Shallice, T. (1988). *From neuropsychology to mental structure.* Cambridge University Press.

Stuss, D. T., & Benson, D. F. (1986). *The frontal lobes.* Raven Press.

Van der Meer, T. G. L. A., Hameleers, M., & Kroon, A. C. (2020). Crafting our own biased media diets: The effects of confirmation, source, and negativity bias on selective attendance to online news. *Mass Communication and Society, 23*(6), 937–967. https://doi.org/10.1080/15205436.2020.1782432

Index

abstract, 11, 16, 22, 32–33, 69, 74, 80, 100, 128, 150
abstract thinking, 11
ACEs, 115, 117
ADHD, 5, 18, 57, 63, 127
adolescence, 105
adulthood, 108
advertising, 146
ageing, 108
anatomical connectivity, 13, 23, 41, 43, 45, 79
anhedonia, 137
animal studies, 28
anterograde, 76, 79, 85
antipsychotic drugs, 57
antisocial personality disorder, 138
attention shifting, 40
attentional control, 8
autism, 133
autobiographical memories, 69
avoidant decision-making style, 92

behavioural flexibility, 40
Brodmann, 28, 30, 34, 38, 50

cell migration, 40, 97
childhood, 99
cognitive bias, 144
cognitive control, vii, 14, 106
cognitive development, 112
cognitive flexibility, 3
computational neuroscience, 83
confirmation bias, 145
constructive cognitive processing, 122
contention scheduling, 62, 148
cool and hot executive functions, 11
cortisol, 113
cytoarchitecture, 38

decision-making deficit, 127
deep brain stimulation, 26
degeneration, 55
delusions, 58, 130, 132

dependent decision makers, 88
dependent decision-making style, 92
depression, 135
depressive realism, 136
development, 97
disorders, viii, 1, 27, 29, 63, 84, 98, 116–118, 128, 137, 139
divergence, 80
dopamine, 57
dopaminergic cells, 55
dynamic filtering theory, 17
dysfunction, 2, 27, 34, 110, 127, 131, 138–139

education, 102
EEG, 83
encoding, 70
episodic buffer, 73
episodic memories, 69
excitatation, 54
executive control system, 73
experiential system, 148

fear response, 40
Flanker task, 129
fMRI, 15
fractation of memory, 67
free will, 147

General Decision-Making Style questionnaire, 88
genetics, 40, 150
goal directed, 4, 33
gross anatomy, 2

hallucinations, 58, 130, 132, 139
hierarchical organisation, 12, 46
hippocampus, 56, 61, 75, 113, 116, 118
hormone changes, 105
HPA axis, 113

impulsivity, 107, 138
individual differences, 94

information processing, 10
inhibition, 54
inhibitory control, 3–4, 11, 18, 106
integrative association area, 32
integrative theory, 17
intuitive decision makers, 88
intuitive decision-making style, 91
Iowa gambling task, 128

lamination, 40
lesion studies, 26
levels of processing model, 74
lobotomy, 25
localisation of function, 31
long-term memory, 68

maturity, 31
memory, 66
model of working memory., 6
multi-store model, 72
myelination, 101, 105

neurogenesis, 50
neuroplasticity, 48
neurotransmitter systems, 55
neurotransmitters, 54
newborn, 98
noradrenaline, 57

parental roles, 102
Parkinson's disease, 55, 63, 127
patient HM, 75
persuasion, 146
Phineas Gage, 24
phonological loop, 73
phylogenetic development, 30
post-traumatic growth, 121
prefrontal cortex, 22
prefrontal cortex function, 31
prenatal, 97
prolonged development, 10, 31, 105–106, 115, 118
psychopathic traits, 117, 138

rational decision makers, 88
rational system, 148

reciprocity, 45
resolution, 45
retrograde, 75, 79–80
risk perception, 109
risky decision-making, 105
risky decisions, 121
rule switching, 40

schizophrenia, 8, 18, 25, 58, 84, 116–117, 123, 127, 130–131, 133, 139
semantic memories, 68
sensory memory, 68
serotonin, 55, 63
short-term memory, 68
signalling, 54
social and moral development, 101
social approval, 107
social media, 145
spontaneous decision-making style, 92
stress, 119
stress response, 114
Stroop effect, 12
substance use, 133
suicide, 137
supervisory attentional system, 63
synapse, 58
synaptic plasticity, 114
synaptic pruning, 50, 101
synaptogenesis, 101
systematic decision-making style, 91

task switching, 8, 16, 18, 101
temporal direction, 69
top-down processing, 12–13, 17
topographic maps, 41, 50
topographic organisation, 41
trauma, 117

visual cortex, 13, 31, 34, 50
visuospatial sketchpad, 73

working memory, 3, 6, 10, 15–16, 27, 40, 58, 63, 73, 75–76, 101, 105–106, 109–110, 117, 123

For EU product safety concerns, contact us at Calle de José Abascal, 56–1°,
28003 Madrid, Spain or eugpsr@cambridge.org.

www.ingramcontent.com/pod-product-compliance
Ingram Content Group UK Ltd.
Pitfield, Milton Keynes, MK11 3LW, UK
UKHW021914211125
465270UK00005B/72